AF552504

ENCYCLOPAEDIA OF ENGINEERING PHYSICS - I

AN INTRODUCTION TO ENGINEERING PHYSICS

By

Dr. Shalender Singh

DISCOVERY PUBLISHING HOUSE PVT. LTD.
NEW DELHI-110 002

Reprinted - 2019

First Published - 2009

ISBN: 978-81-8356-349-9 (Set)

ISBN: 978-81-8356-412-0

An Introduction to Engineering Physics

Published by:

DISCOVERY PUBLISHING HOUSE PVT. LTD.
4383/4B, Ansari Road, Darya Ganj
New Delhi-110 002 (India)
Phone: +91-11-23279245, 43596064-65
Fax: +91-11-23253475
E-mail: discoverybooksindia@gmail.com
discoverypublishinghouse@gmail.com
web: www.discoverypublishinggroup.com

Printed at:
Infinity Imaging Systems
Delhi

Preface

The present book "Encyclopaedia of Engineering Physics" meant for engineering students. The present book is an attempt to fulfil the need of all engineering students of U.P.T.U. as well as for engineering students of other states. It covers the complete syllabus of physics prescribed by Technical Universities. The treatment given in simple, lucid and comprehensive. The language used is simple.

Though nothing can be claimed as original but the subject matter has been arranged in my own style.

Suggestion for improvement of the book will be greatly acknowledged.

Author

CONTENTS

1
Relativistic Mechanics

INTRODUCTION

Few, indeed, if any, have so completely shaken the world of science, since *Newton*, by the sheer boldness and profoundness, of their ideas. He literally created an upheavel by the publication, in quick succession, in the year 1905, two epoch-making papers, on the concept of the photon and on the *Electrodynamics of moving bodies* respectively, with yet another on *the mathematical analysis of Brownian motion* thrown in, in between, try far the most revolutionary of them is the one on the electrodynamics of moving bodies which demolishes at the stroke some of the most cherished and supposedly infallible laws and concepts and gives the breath-takingly new idea of the relativity of space and time We have here no revolutionary act but the natural continuation of a line that can be traced through centuries. The abandonment of a certain concept connected with space, time and motion hitherto treated as fundamental must not be regarded as arbitrary but only conditioned by observed facts. In the person of *Albert Einstein*, appear on the firmament among the all-time great that includes *Newton* and *Maxwell.*

The law of the constant velocity of light in empty space which has been confirmed by the development of electrodynamics and optics and the equal legitimacy of all inertial systems (special theory of relativity) which was proved in a particularly incisive manner by Michelson's famous experiment, between them made it necessary, to begin with that the concept of time should be made relative, each inertial system being given its own special time.... It is, in general, one of the essential features of the theory of relativity that it is at pains to work out the relations between general concepts and empirical facts more precisely. The fundamental principle here is that the justification for a physical concept lies exclusively in its clear and unambiguous relation to facts that can be experienced."

CONSERVATION OF RELATIVISTIC ENERGY

The relativistic energy, as we know, is given by the relation $E = mc^2 = \gamma/m_oc^2$, where the value of m, equal to fino, has been deduced on the validity of the law of conservation of relativistic momentum. It follows, therefore, that relativistic energy too is subject to a law of conservation similar to that of momentum. Let us, however, deduce the law in a more direct manner as explained below.

Imagine a reference frame S in which two given particles, with momenta p_1 and p_2 and energies E_1 and E_2 respectively, undergo a collision. If their respective momenta after the collision be p_3 and p_4 and their energies E_3 and E_4, we nave, in accordance with the law of conservation of momentum,

$$P_1 + P_2 = P_3 + P_4.$$

Or, if we consider the x-components of their momenta only, we have

$$P_1 + P_{2x} + = P_{3X} + P_{4x}$$

In another frame S' moving with velocity v relative to S along the axis of x, the components of the initial and final momenta of the particles along the x'-axis will respectively be p_{1x}', p_{2x}', p_{3x}', and p_{4x}', and their energies before and after the collision will be E_1, E_2 and E_1', E_2'.

Since the law of conservation of momentum (being a Lorentz invariant) holds good in this frame too, we have $p_{1x}' + p_{2x}' = p_{3x}{}^{+} p_{4x}$

Or, $\gamma(p_{1x} - vE_1/c^2) + \gamma(p_{2x} - vE_2/c^2) = \gamma(p_{3x} - vE_3/c^2) + \gamma(p_{4x} - vE_4/c^2)$

Or, $(p_{1x} - p_{2x}) - v/c^2\ (E_1 - E_2) = (p_{3x} - p_{4x}) - v/c^2\ (E_3 - E_4)$.

Since, in accordance with the law of conservation of momentum,

$$(p_{1X} + p_{2X}) = (p_{3X} + p_{4X}),$$

We have $\quad E_1 + E_2 = E_3 + E_4,$

i.e., the energy of the two particles is also conserved in frame S.

It may similarly be shown that $E_1' + E_2' = E_3' + E_3'$ or that *the energy of the two particles is also conserved in frame S'.*

What is true for two particles is also equally true for any number of them. We thus see that *the law of conservation of energy too, like the law of conservation of momentum, is a Lorentz invariant, i.e., holds good in all (inertial) frames of reference.*

RELATION BETWEEN RELATIVISTIC MOMENTUM AND ENERGY

A simple relation between relativistic momentum (p) and relativistic energy (E) of a particle of rest mass m_o is easily obtained from the expressions for p and E, viz., $p = \gamma m_o v$ and $E = \gamma m_o c^2$. For, clearly, $\gamma m_o = E/c^2$. So that, $p = Ev/c^2$,

which enables us to determine the velocity of the particle from the values of its momentum and energy, since, clearly, $v = pc^2/E$/

Another *important relation* between p and E may be obtained as follows:

Since $p = \gamma m_o v$, we have $p^2 = \gamma^2 m_o v^2$.

And since $\gamma = 1/\sqrt{1 - v^2/c^2}$,

we have $\gamma^2 = 1/(1 - v^2/c^2) = c^2/(c^2 - v^2)$,

whence, $\gamma^2 v^2 = \gamma^2 c^2 - c^2$.

Substituting this value of $\gamma^2 v^2$ in the expression for p^2 above, we obtain $p^2 = (\gamma^2 c^2 - c^2)m_o^2$, multiplying which by c^2 we get

$$p^2c^2 = m_o^2\gamma^2c^4 - m_o^2c^1.$$

Or, $$m_o^2\gamma^2c^4 = p^2c^2 + m_o^2c^4.$$

Now $\gamma m_o c^2 = E$, total energy of the particle in question. So that,

$$E^2 = p^2c^2 + m_o^2c^4$$

Or, $$E^2 - p^2c^2 + m_o^2c^1$$

an expression connecting relativistic energy and momentum.

Since m_o and c are both Lorentz invariants, $m_o^2c^1$ is a constant and therefore, the quantity $E^2 - p^2c^2$ too is a *Lorentz invariant* and its value remains unaltered in any inertial frame of reference.

EQUIVALENCE OF MASS AND ENERGY

The correct way to state the second law of motion, in accordance with the Special theory of relativity, is to define force as the time rate *of change of momentum, i.e.*, as $F = d/dt\ (mv)$, since not only does it reduce to the classical second law of motion at low values of velocity but also, taken together with the third law, ensures the validity of the law of conservation of momentum in a closed system even under relativistic conditions.

This is obviously not the same definition as, F = ma, since

$$\frac{d}{dt}(mv) = m\frac{dv}{dt} + v\frac{dm}{dt} = ma + v\frac{dm}{dt},$$

and dm/dt ∞ 0 if v varies with time.

Now, in mechanics (Newtonian as well as Relativistic), the kinetic energy of a moving body at a given velocity v is equal to the work done in making the body move from rest and attain this velocity. Thus, if F be the component of the force applied to a body in the direction of displacement, the work done or kinetic energy imparted to the body when it is displaced through a distance dS, *i.e.*, dE_k = F.dS Therefore, work done or kinetic energy imparted to it during the whole displacement S until it acquires velocity v is given by

$$K.E. = E_k = \int dE_k = \int_0^S F.ds = \int_0^S \frac{d}{dt}(mv)\, dS.$$

Now, $\frac{d}{dt}(mv)\, ds = \frac{dS}{dt} d(mv) = v d(mv)$ $[\because dS/dt = v.$

Since the particle starts from rest (*i.e.*, v = 0) and finally acquires velocity v, we have

$$K.E. = E_k = \int_0^v v d(mv) = \int_0^V v d\left(\frac{m_0 v}{\sqrt{1 - v^2/c^2}}\right) \quad \left[\because m = \frac{m_0 v}{\sqrt{1 - v^2/c^2}}\right.$$

Integrating this by parts, we have

$$E_k = \frac{m_0 v^2}{\sqrt{1 - v^2/c^2}} - m_0 \int_0^V \frac{v dv}{\sqrt{1 - v^2/c^2}}$$

$$= \frac{m_0 v^2}{\sqrt{1 - v^2/c^2}} + m_0 c^2 \left[\sqrt{1 - v^2/c^2}\right]_0^V$$

Or, $$E_k = \frac{m_0 c^2}{\sqrt{1 - v^2/c^2}} - m_0 c^2 = mc^2 - m_0 c^2 = (m - m_0)c^2 = \Delta mc^2,$$

where Δm is the relativistic increase in mass with increase in velocity.

Thus, the K.E. of the body is the product of the increase in its mass and c^2.

Or, rewriting the equation, we have

$$mc^2 = m_oc^2 + E_k = \frac{m_oc^2}{\sqrt{1 - v^2/c^2}} = \gamma m_oc^2$$

Here, m_oc^2, the energy due to the rest mass of the body, *i.e.*, its energy when at rest with respect to the observer, is called its *rest energy* or *proper energy* E_0 and mc^2, the total energy E possessed by the body.

So that, we have

total energy E(= mc²) = rest energy (=m₀c²) + kinetic energy (E_k) γ = m₀c².

This relation may also be obtained straightaway by Binomial expansion of the mass-velocity relation for values of $v/c \rightarrow 0$. Thus,

$$m = \frac{m_o}{(1 - v^2/c^2)^{1/2}} = m_o(1 - v^2/c^2)^{-1/2}$$

$$= m_o\left(1 + \frac{1}{2}\frac{v^2}{c^2} - ...\right).$$

Or, $m = m_o + 1/2\, m_ov^2/c^2$,

whence, $mc^2 = m_o + 1/2\, mv^2$. Or, $mc^2 = m_oc^2 + E_k = \dfrac{m_oc^2}{\sqrt{1 - v^2/c^2}} = \gamma m_oc^2$.

∴ therefore, kinetic energy is given by the relation,

$$E_k = (m - m_o)\, c^2 = \Delta mc^2.$$

This relativistic expression for K.E., or this mass-energy equation, referred to as *Einstein's mass-energy relation*, is by far the most famous of the equations deduced by Einstein and stands fully verified and confirmed by experiment.

It will be easily seen that in view of the high value of c^2, viz., 9×10^{20}, even a small amount of mass Δm is equivalent to a large amount of energy. The change in the mass (Δm) in cases with which we are concerned in our every day life are, however, much too small for any energy effects to be visible. But, in the case of radioactive substances, for example, the enormous energy, thus produced by the conversion of mass, enables them to eject a and P panicles with such high velocities and over such long periods. Or, again, the estimated decrease in the mass of the mn being 4×10^{12} gm/sec, it radiates out is energy equivalent of $4 \times 10^{12} \times 9 \times 10^{20} = 3.6 \times 10^{23}$ erg/sec.

It will thus be noted that the classical expression $1/2\ mv^2$ for K.E. does give the true K.E. of the body even if we substitute the relativistic values of m and v in it.

On the other hand, the expression $E_k = mc^2 = m_oc^2$ reduces to the classical expression $E_k = 1/2\ mv^2$ for values of $v << c$. For, we have

$$E_k = \frac{m_o}{\sqrt{1 - v^2/c^2}} - m_oc^2 = m_oc^2\ (1 - v^2c^2)^{-1/2} - m_oc^2$$

$$= m_oc^2 \left(1 - \frac{1}{2}\frac{v^2}{c^2} ......\right) - m_oc^2 = \frac{1}{2}m_ov^2,$$

{The higher terms in v/c being neglible when v/c << 1.}

a result in conformity with classical physics.

We thus see that the relativistic expressions for *length, time, mass* and *energy*, all obey the same corresponding principle, viz., that they get reduced to their classical forms for small values of v when $v/c \rightarrow 0$.

This shows that the former are the really true expressions and the latter, merely approximations to them, valid only when v is small. Since, however, we seldom come across values of v comparable with c, except in atomic physics, the classical expressions remain true enough for our manifold practical purposes.

Importance of mass-energy equation : It will be seen from relation II above, viz., $m = m_o + 1/2\ m_ov^2/c^2$, *that the total menial-mass of a particle moving with velocity v relative to the observer, is the sum of (i) its rest-mass* m_o, *and (ii) an additional mass (or inertia) equal to* $E_k\ c^2$.

This suggests at once that the addition of energy to a system increases the inertial mass of the system. In plain and simple terms, it means that *mass may appear as energy and energy as mass*. So that, the scope of the law of conservation of energy may be widened into what may be called the *law of conservation of mass-energy*, meaning thereby that, *in any interaction, it is neither mass alone nor energy alone that remains conserved but mass, inclusive of energy in terms of mass, or energy inclusive of mass in terms of energy, that is realty conserved.*

Here are some illustrative examples:

(i) When a *positron* (an elementary particle, having the same mass as an electron and a charge equal and opposite to it, *i.e.*,+ e) and an *electron* come together, they *annihilate* one another and an equivalent amount of emerge in the form of a pair of γ-ray *phutuns* (of energy at least 1.02 Afev) appears in their place.

The reverse of this phenomenon of annihilation of matter is that of lair production, where a *photon* of energy at least 1.02 Mev gives rise 10 a *pair of an electron* and *σ positron* in the intense electric field near the nucleus of an atom. Here is what is called *materialisation of energy, i.e., conversion of energy into matter.*

Another example of pair production is the production of a *proionaniprtion pair*, requiring a minimum of energy 1.08 *Giga electron* volts.

The laws of conservation of momentum, mass-energy and of charge hold good in both these processcs.

(ii) When a uranium nucleus is split up, the decrease in its total rest mass appears in the equivalent amount of kinetic energy of its fragments. Thus, suppose it is split up into equal fragments, moving with velocity t, so that the mass of each is no longer its rest mass m_0 but, say, m_v, which is greater than m_0. The sum of the masses of the two fragments is, therefore, $2m_v = M$, say, which is greater than their rest mass $2m_0$. The kinetic energy liberated by the splitting nucleus, or the energy due to conversion of $(M - 2m_0)$ of matter is thus $E_k = (M - 2m_0)c^2$.

As can be easily seen, it is thus possible to calculate the amount of energy that is released due to *fission* in an atomic bomb, since both M and m_0 are known. Distressingly enough, it was for just suggesting this method of calculating the energy released by an atomic bomb that Einstein had been, unthinkingly and uncharitably, dubbed the world over as the 'father of the atomic bomb'.

(iii) In the process of fusion of nuclei, the total mass of the fusing nuclei is somewhat in excess of the nuclei produced as a result thereof, the difference being convened into energy in accordance with the mass-energy relation In fact, such conversion of matter is the main source of energy of the sun and the stars, which obtain their enormous energy chiefly from what is called the *'burning' of protons into helium gas, i.e., the fusion of protons to form helium atoms.* The change in mass during the formation of one helium atom is given by Δm = *(mass of 4 protons + mass of 2 electrons) – (mass of one helium atom). So that,*

$$\Delta m = (4 \times 1.6725 \times 10^{-24} + 2 \times 0.911 \times 10^{-24}) - (6.647 \times 10^{-24} \text{ gm.}$$

Hence *energy released pechelium atom produced* $= \Delta mc^2$

$$= 0.045 \times 10^{-24} \times (3 \times 10^{10})^2 \approx 25 \text{ Mev}$$

$$[\because 1 \text{ ev} = 1.6 \times 10^{12} \text{ erg.}$$

Again, when a number of particles, called *nucleons, i.e., protons* and *neutrons*, combine to form a nucleus, there is a loss of mass Δm, *i.e.,* the mass of the nucleus is this much less than that of its constituent protons and neutrons. *The energy-equivalent of this mass-loss, i.e.,* $E = \Delta mc^2$ is called the *binding energy* of the nucleus.

It may be taken as a general rule that *an increase in mass means an increase in energy and vice versa,* though in most everyday cases, these changes in mass or energy are much too small to be discernible. Thus, a gas or a spring should, in accordance with the theory, record a greater mass in the compressed state, the increase in mass, obviously, being the work done to compress it divided by c^2. Similarly, a body when hot (and, therefore, with greater internal energy) should have a somewhat higher mass than when it is cold, and soon.

TRANSFORMATION OF RELATIVISTIC MOMENTUM AND ENERGY

Consider a particle of rest mass m_0, moving in a reference frame S. Its momentum in this frame is given by $p = \gamma m_0 dr/dt$ and its energy by $E = \gamma m_0 c^2$, where dr/dt is its velocity.

As we know, an interval of time dt, as noted on a clock in the reference frame S is related to the *proper interval of time* $d\tau$, as noted on a clock carried by the particle itself, as $dt = d\tau dt = d\tau/\sqrt{1 - v^2/c^2}$, whence $dt = d\tau/ = 1\sqrt{1 - v^2/c^2} = \gamma$.

So that, if p_x, p_y, and pa be the components of the momentum of the particle along the three coordinate axes, we have

$$p_x = \gamma m_o \frac{dx}{dt} = m_o \frac{dx}{dt}.\frac{dt}{d\tau} = m_o \frac{dx}{d\tau} \text{ and, similarly } p_y = m_o \frac{dy}{d\tau}$$

$$p_z = m_o \frac{dz}{d\tau} \text{ and } E = m_o c^2 \frac{dt}{d\tau} \text{ or } \frac{E}{c^2} = m_o \frac{dt}{d\tau}.$$

Since both m_o and τ are Lorentz invariants, the quantities p_x, p_y and p_z, get transformed in exactly the same manner as x, y, z and t in a reference frame S', moving with velocity v relative to S along the axis of x. And, we therefore have

$$p_x' = \gamma (p_x - vE/c^2),\ p_y' = p_y,\ p_z' = p_z$$

and $$E'/c^2 = \gamma \left(\frac{E}{c^2} - \frac{v}{c^2} p_x \right)$$

or $$E' = \gamma(E - vp_x).$$

The inverse Lorentz transformation will thus obviously be

$$p_x = \gamma(p_x' + vE'/c^2),$$

$$p_y = p_y' = p_z = p_z'$$

and $E = (E' + vp_x')$.

ATOMIC MASS UNIT

The basic mass unit used in atomic physics is called the atomic *mass unit* (*amu*, for short), which is $1/12^{th}$ of the rest mass of *carbon twelve* (C^{12}), *very nearly equal to the mass of a proton. Thus, 1 amu = 1.66 × 10⁻²⁷ kg.*

Since, energy associated with a rest mass m_o is $E = m_o c^2$, we have *1 amu, in terms of energy*

$= 1.66 \times 10^{-7}$ kg $\times (3 \times 10^6)^2 m^2/sec^2$

$= 1.49 \times 10^{-10}$ *joules.*

$= 1.49 \times 10^{-10} \times 10^7 = 1.49 \times 10^{-3}$ ergs.

And, since 1.60×10^{-19} joules = 1 eV (electron volt), we have

$$1 \text{ amu} = \frac{1.49 \times 10^{-10}}{1.60 \times 10^{-19}} \text{ eV} = 9.31 \times 10^8 \text{ eV} = \frac{9.31 \times 10^3}{10^6} = 931 \text{meV}$$

(mega electron volt).

MASS VELOCITY, MOMENTUM AND ENERGY OF PARTICLES OF ZERO REST MASS

We have just seen above that the energy of a panicle of rest mass m_0 and momentum p, is given by

$$E = (p^2c^2 + m_0^2c^4)^{1/2}$$

If therefore, we have a particle of rest mass zero, like a *photon* or a *nutrino, clearly* $m_0 = 0$ and $\therefore$ $E = pc$, whence, $p = E/c$.

And since, as we have seen p is also equal to Ev/c^2 we have

$$Ev/c^2 = E/c, \text{ whence, } \quad v = c,$$

i.e., the velocity of a particle of zero rest mass is always c, the speed of light, and is, therefore, invariant and so is its rest mass (zero), in all frames of reference.

And, its energy is $\qquad E = mc^2$

where m is the mass equivalent of energy E, equal to E/c^2

To summarise then,

(i) the mass of a particle of zero rest mass, i.e., $m = E/c^2$.

(ii) its velocity is c, the same as the speed of light in free space.

(iii) its momentum $p = mc = E/c$ *and*

(iv) its energy $E = mc^2 = pc$.

In case the particle be & photon of frequency v. We have $E = hv$, where h is the well known *Planck's constant.*

So that, for a photon, mass $\quad m = E/c^2 = hv/c^2$ and

momentum $\quad p = E/c = hv/c.$

ULTIMATE SPEED OF A MATERIAL PARTICLE

In classical mechanics, there being no upper limit to velocity it is possible that as a particle is given more and more acceleration, its speed may go on increasing progressively and may well become greater than c,—in fact, it may have any velocity whatever. This is *firmly* denied by the theory of relativity. It may legitimately be asked, therefore, as to what will happen if the particle is continually accelerated. Certainly, its velocity v goes on increasing and hence also its *mass* in accordance with the mass-velocity relation $m = m_0/\sqrt{1 - v^2/c^2} = \gamma m_0$. But as v approaches c, $v^2/c^2 \rightarrow 1$ and therefore $\sqrt{1 - v^2/c^2} \rightarrow 0$ and the mass of the particle

(m) → ∞, as shown graphically in Fig. 1.1, from which it is clear that for velocities right up to c/2, the increase in mass from the value of the rest or inertial mass m_o is quite inappreciable. Beyond this, however, the mass increases relatively more rapidly and *tends to become infinite* as v → c. whereas momentum goes on increasing the velocity remains constant, indicating that the accelerating force applied only increases the mass and hence the momentum but not velocity.

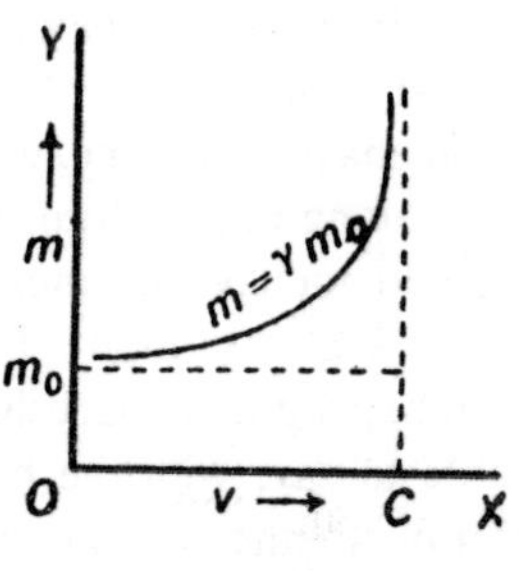

Fig. 1.1

The idea of infinite mass, however, makes no sense for a number of reasons. Thus, for example, (i) *it would require an infinite force to accelerate the particle to the speed of light c, (ii) due to Lorentz-Fitzgerald contraction, its length in its direction of motion must be zero, (iii) its volume too must, therefore, be zero, and (iv) it must exert an infinite gravitational pull on all other bodies in the universe etc., etc.*

So that, discarding the notion of an infinite mass, we interpret the above relationship to mean that *no material body can equal or exceed the speed of light in free space, viz., 3 × 10^{10} cm or 3 × 10^8m/sec.*

Now, it may perhaps be cited in apparent refutation of this statement, that a *photon*, in free space, travels with the free-space speed c of light. So, of course, it does, but then a photon is not a material particle since it possesses no rest mass. Also, there are, what are called phase velocities, which may even exceed the free-space speed c of light, but these are, again, no material particles but only abstract mathematical functions. Finally, some of the atomic particles may travel in media like air, glass or water with velocities higher than that of *light in those media but by no means higher than its speed in free space.*

Thus, we come to the conclusion that the ultimate *speed for any material particle is the speed of light in free-space, which it can never attain or exceed.*

EINSTEIN'S CONCEPT OF RELATIVITY

When the experiments in search of the ether drift failed, it began to be increasingly realised that there was no such thing as an *absolute* or *privileged* frame of reference and that the basic laws of physics took the same form in all inertial frames of reference. The implications of this

Galilian invariance principle were emphasised by the French mathematical physicist *Henri Poihcar'e* when he stated that '... the laws of physical phenomena are the same, whether for an observer fixed or for an observer carried along in a uniform movement of translation, so that we have not and could not have any means of discerning whether or not we are carried along in such a motion'. This simply means that if we are drifting with uniform speed in a spaceship, with all the windows closed, we shall not be able to say, with the help of any experiments we might choose to perform, whether we are at rest or in motion. If, however, we look out of a window, we shall be able to say merely that we are in motion with respect to the fixed stars but not whether we or the stars are actually in motion.

How near, indeed, had Poincare thus come to expounding the theory of relativity and yet how far he actually was from it. For, instead of grasping the implications of the failure of all ether-drift experiments, and building up a new theory on its basis, discarding older notions of space and time, he concerned himself with trying to somehow save the old classical theory by suitable adjustments and modifications in it.

The real import of the negative results of the ether-drift experiments was clearly seen and understood by *Einstein*. For, discussing the reciprocal electrodynamic action of a magnet and a conductor where 'the experimentally observable phenomenon depends only on the relative motion of the conductor and the magnet...', he goes on to say:

'Examples of this sort, together with the unsuccessful attempts to discover any motion of the earth relative to the 'light medium', suggest that the phenomena of electrodynamics as well as of mechanics possess no properties corresponding to the idea of absolute rest.

They suggest rather that.... the same laws of electrodynamics and optics will be valid for all frames of reference for which the equations of mechanics hold good. We will raise this conjecture (the purport of which will hereafter be called the *Principle of Relativity*) to the status of a postulate, and also introduce another postulate which is only apparently irreconcilable with the former, namely, that light is always propagated in empty space with a definite velocity c which is independent of the state of motion of the emitting body .... The introduction of a "luminiferous ether" will prove to be superfluous in as much as the view here to be developed will not require an 'absolutely stationary space' provided with special properties, ...

Thus, from the negative results of the ether-drift experiments and from his own reasoning, *Einstein* felt fully convinced that there was no such thing as an *absolute* or *fixed* frame of reference. He examined the physical consequences of the absence of such a frame of reference and had the boldness to break away from old and traditional concepts of space and time. He knit his conclusions and revolutionary ideas into a cogent theory which he announced to an unsuspecting world in the year 1905 as his *Special theory of relativity*. And, ten years later, in 1915, followed the second and the more complex and difficult part of it in the form of the General theory of relativity. The former deals with problems associated with unaccelerated frames of reference, i.e., those which move with uniform relative velocity with respect to one another and the latter with those associated with accelerated ones.

We shall concern ourselves here mainly with only the Special theory of relativity which is not only comparatively simpler but has also produced the most profound effect on the entire field of physics.

SPECIAL THEORY OF RELATIVITY

The two basic postulates on which the theory rests may be stated in a precise form thus:

1. The laws of physics all take the same identical form for all frames of reference in uniform relative motion, *i.e.*, for all the inertial frames of reference.

   This, it will be readily seen, is a direct consequence of the absence of an absolute or fixed frame of reference. For, if the laws of physics were to take on different forms in different frames of reference, it would be easily inferred from these differences as to which of them are at rest in space and which in motion. But, as we have seen, such distinction between the state of- rest and of uniform motion is precluded by the absence of a universal frame of reference. The first postulate is thus merely a generalised statement of this observed fact.

2. The velocity of light in free space is the same (c) relative to any inertial frame of reference, i.e., it is invariant to transformation from one inertial frame to another and has thus the same value (3.0×10^{10} cm/or 3×10^{5} m/sec) for all observers irrespective of their state of motion.

This postulate is clearly a statement of the result of *Michelson—Morley* experiment.

The two implications of the above postulates are immediately obvious:

(i) Velocity being not invariant to Galilian transformation, it follows that if c be the velocity of light in frame S, that in frame S', moving relative to 5 with velocity v, must be c' = c − v. In accordance with the second postulate of the special theory of relativity, however, we have c' = c since c must always have the same value irrespective of the state of motion of the frames, of reference.

The *new postulate, is thus dearly at variance with Galilian transformation.*

(ii) Again, suppose a flash of light is emitted from the origin when points O and O' of the two frames of reference S and S' Fig. 1.2 just cross each other. Suppose further that it is possible for the two observers in the two frames of reference to see the light or rather the *wave front* (a surface of equal phase) spreading out into space. Then, in view of the constancy of the value of c, irrespective of the motion of the two frames of reference, each of the two observers at O and O' respectively will, even after drifting apart, claim to be at the centre of the spreading spherical wave front of light, as shown in Fig. 1.3, thinking that the other has moved away from the centre of the sphere.

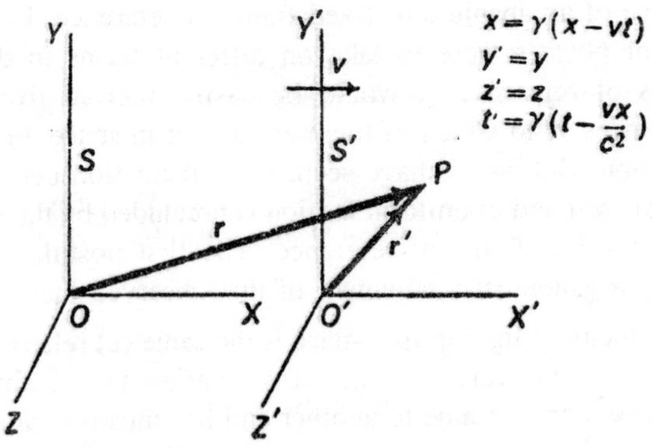

Fig. 1.2

As can be readily seen, this will not be the case if, instead of being in free space, our observers were to be in boats in water and if a stone were dropped (instead of a flash of light emitted) when they just crossed each other. The ripple pattern formed will appear to be different to each observer and they will be able to say from the very shape of it whether or not they are in motion or which one of them is in motion and which, stationary.

This difference between the two cases, it may be carefully noted, is due to the fact that *whereas water functions as a frame of reference by itself, 'space does not, and, again, whereas the speed of the waves in water varies with the motion of the observer, the speed of the waves in space (i.e., of light or electromagnetic waves) is quite independent of it.*

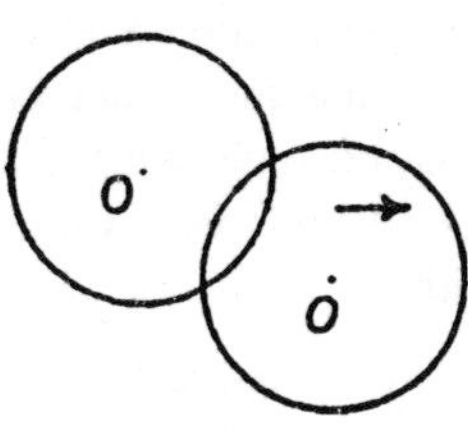

Fig. 1.3

LORENTZ COORDINATE TRANSFORMATION (or, simply, LORENTZ TRANSFORMATION)

As just mentioned the Galilian transformation is quite at variance with the velocity of light in free space being independent of the motion of the observer or the source of light. We, therefore, look for a transformation which is consistent with, this new concept.

As we have seen under the article referred to, if a flash of light be emitted as the two observers just cross each other in the two frames of reference Fig. 1.2, each thinks himself to be at the centre of a uniformly expanding spherical wave front Fig. 1.3. Clearly, *equation of the wave front, as seen by the observer at 0 in reference frame S, is*

$$x^2 + y^2 + z^2 = c^2t^2, \qquad ...(i)$$

And *equation of the wavefront, as seen by the observer at 0' in reference frame S'*, is

$$x^2 + y^2 + z^2 = c^2t^2, \qquad ...(ii)$$

where c, the velocity of light is the same in either frame of reference.

In order, therefore, that both the observers may seem to be at the centre of the same expanding wave front, we should have relation (i) = relation (ii).

Now, *Galilian transformation* gives $x' = x - vt$, $y^2 = y$, $z' = z$ and $t' = t$. So that, substituting for x', y', z' and t' in relation (ii) above, we obtain

$$x^2 - 2xvt + v^2t^2 + y^2 + z^2 = c^2t^2, \quad \text{...(iii)}$$

which is surely not the same as relation (i), showing once again, that *the Galilian transformation fails if the constancy of the value of c be assumed.*

This means clearly that in the new transformation we are looking for, the relation $x' = x - vt$ no longer holds good. Since, however, the relation is quite in accord with classical mechanics, the new trans- formation must be such that the x-coordinate, in accordance with it, also reduces itself to this very value, when the relative velocity v is small compared with c.

The simplest relation for the x-coordinate, in accordance with the new transformation, can be

$$x' = \gamma(x - vt), \quad \text{...(iv)}$$

where γ is independent of the coordinates x or t, but may vary with v.

This relation has the merit of being a *linear* one in x and t, providing for a uniformly expanding wave front and admitting of one and only one interpretation in the system S' of an observation made from system S. A. quadratic equation, on the other hand, will obviously give two; and a higher order equation, even more than two interpretations, which is simply inadmissible.

In plain and simple language, the equation $x' = \gamma(x - vt)$ will automatically give only one value of x' for a given value of x, without our having to impose any further conditions.

Considering the inverse transformation, connecting measurements in frame S with those made in frame S', we shall have only the sign of v reversed; so that,

$$x = \gamma(x' + vt), \quad \text{...(v)}$$

the constant of proportionality γ remaining the same in either case, since the first postulate of the theory rules out any preferred frame of reference.

And, since the relative motion of S and S' is confined to the x-direction, we have $y' = y$ and $z' = z$.

It may be noted that t' *can no longer be taken to be equal to t, or* else relations (iv) and (v) cannot both be true.

Now, substituting in relation (v) the value of x', as given by relation (iv), we have

$$x' = \gamma[\gamma(x - vt) + vt'] = \gamma^2 (x - vt) + \gamma vt'.$$

Or, $\gamma vt' = x - \gamma^2 (x - vt) = x - \gamma^2 x + \gamma^2 vt = \gamma^2 vt + x(1 - \gamma^2)$,

whence, $$t' = \gamma t + x\left(\frac{1-\gamma^2}{\gamma v}\right) \qquad \text{...(vi)}$$

In order to determine the value of γ, we note that the reference frame S' moves with relative velocity v in the + x-direction with respect to reference frame S and that the flash of light is emitted when the two observers at O and O' are opposite each other, *i.e.*, at the time the flash is emitted, x = x' = 0 and also t = t' = 0. Since the velocity of light is the same c in the two frames of reference, we have

x-coordinate of the light pulse in frame S, given by x = ct. ...(vii)

and x'-coordinate of the light pulse in frame S', given by x' = ct' ...(viii)

Substituting in relation (viii) the values of x' and t' from relations (iv) and (vi) respectively, we have

$$\gamma(x - vt) = c\gamma t + cx\left(\frac{1-\gamma^2}{\gamma v}\right)$$

whence, solving for x, we obtain $x = ct\left[\dfrac{1 + v/c}{1 - (c/v)(1/\gamma^2 - 1)}\right]$.

since x = ct, as shown in relation (vii) above, we have

$$\frac{1 + v/c}{1 - (c/v)(1/\gamma^2 - 1)} = 1.$$

Or, $$1 + \frac{v}{c} = 1 - \left(\frac{c}{v}\right)\left(\frac{1}{\gamma^2} - 1\right),$$

whence, $$\gamma = \frac{1}{\sqrt{1 - v^2/c^2}}$$

The factor v/c is quite often represented by β; so that,

$$\gamma = \frac{1}{\sqrt{1 - v^2/c^2}} = \frac{1}{\sqrt{1 - \beta^2}}$$

Substituting this value of γ in relations (iv) and (vi) above, we have

$$x' = \frac{dx - vt}{1 - v^2/c^2} = \gamma\,(x - vt) \quad ...(ix)$$

and $$t' = \gamma\left(t - \frac{vx}{c^2}\right) = \frac{t - vx/c^2}{\sqrt{1 - v^2/c^2}} \quad ...(x)$$

And, since, as already mentioned, S' moves along the + x direction relative to S, we have

$$y' = y \;...(xi) \text{ and } z' = z. \;...(xii)$$

These values of x', y', z' and t', when substituted in relation (ii) above, give relation (i), precisely as it should be if the two observers are to seem to be at the centre of the same spherical wave front.

This *relativistic transformation* is referred to as *Lorentz transformation*, after *H.A. Lorentz*, a Dutch physicist, because it was be who first showed that this transformation was the only one under which the laws of electricity and magnetism have the same form in all reference frames in relative motion (*i.e.*, in all *inertial* frames), though it was *Einstein*, of course, who showed its wider significance as being the only proper transformation for *all* types of measurements, electromagnetic or otherwise.

Equations (ix), (x), (xi) and (xii) are, therefore, referred to as *Lorentz transformation equations, or Lorent? coordinate transformation equations.* The Galilian and Lorentz (or Relativistic) transformations for an observer in reference frame S', moving with uniform velocity v relative to frame S in the + x-*direction*, are arranged side by side below, so that the differences between the two may become apparent at a glance:

Galilian transformations	Lorentz (or Rclatmstic) transformations
$x' = x - vt$	$x' = \dfrac{x - vt}{\sqrt{1 - v^2/c^2}} = \gamma(x - vt)$
$y' = y$	$y' = y$
$z' = z$	$z' = z$
$t' = t$	$t' = \dfrac{t - vx/c^2}{\sqrt{1 - v^2/c^2}} = \gamma\left(t - \dfrac{vx}{c^2}\right)$

N.B. : If v/c be small, as it most often would be, we may take only the first two terms of the Binomial expansion of $\frac{1}{\sqrt{1-v^2/c^2}}$ or $\left(1-\frac{v^2}{c^2}\right)^{-1/2}$ namely, $\left(1-\frac{1}{2}\frac{v^2}{c^2}\right)$ as the value of γ. This approximation is valid enough even if v/c has a value up to about 0.5.

As will be readily seen, when $v << c$, *i.e.*, when $v/c \rightarrow 0$, we have $\gamma = 1/\sqrt{1-v^2/c^2} \approx 1$ and Lorentz (or Relativistic) transformations reduce to Galilian ones. So that, Galilian or Newtonian physics is just a particular case of Relativistic physics. In fact, v for the earth (18.5 mi/sec) being only 1/10,000th of the value of c, is negligibly small compared with c and the Galilian or Newtonian physics is thus perfectly valid for almost all our practical purposes, the relativistic effects manifesting themselves only at very high values of v, when they are comparable with c, as in the case of electrons, protons, mesons etc.

We have, in our discussion above, assumed frame S' to be moving with velocity v in the positive direction of x relative to frame S, As seen from frame S', therefore, frame S would appear to be moving in the negative x direction, *i.e.*, with velocity –v. Consequently, the *Inverse Lorentz transformation*, which converts measurements made in frame S' into those in frame S, may be obtained from Lorentz transformations by simply changing the sign of v from positive to negative. We shall then have

$$x = \frac{x' + vt'}{\sqrt{1-v^2/c^2}} = \gamma(x' - vt'),\ y = y',\ z = z'$$

and

$$t = \frac{t' + vx'/c^2}{\sqrt{1-v^2/c^2}} = \gamma\left(t' + \frac{vx'}{c^2}\right).$$

It is thus amply clear by now that *measurements of space and time are by no means absolute but are dependent Upon the relative motion between the observer and the phenomenon observed.*

SEARCH FOR A FUNDAMENTAL FRAME OF REFERENCE

It has already been pointed out in the last that it was on account of Newton's insistence as to the existence of a *fundamental* or *absolute frame of reference, called absolute space*, that the search for it was carried on. This search resulted in the discovery, or rather the invention, of that monstrosity of a medium, *luminiferous* ether, as the following brief account will show.

As we know, *Maxwell* clearly demonstrated, in the year 1864, the interrelationship between electricity, magnetism and light when, from the known properties of electricity and magnetism, he formulated his celebrated *theory of electromagnetic radiation* and gave the well known *equations of electromagnetic field* which, bear his name and are identical with those that represent a wave phenomenon. He thus established the presence of electromagnetic waves in space, travelling with the speed of light. In other words, he *proved light to be an electromagnetic wave.*

Since waves known hitherto (like sound and water waves) all required a material medium (air and water respectively) for their propagation, it was supposed that there must also be a suitable medium to carry these newly found electromagnetic waves which travelled even through empty space between the stars and the earth. This intangible medium, though no one knew what it actually was, came to be referred to as *luminiferous ether*, pervading all space, empty or otherwise.

Further, to enable it to transmit light, a transverse wave motion, the ether had to be a *rigid solid* and, in view of the tremendous velocity of light, it had to have a *large shear modulus* and yet all material objects, like the earth and the planets and stars etc., were to continue in their regular courses through it without encountering the slightest resistance. Utterly incredible as it seems to us today, no one seemed to object to its existence. On the other hand, it was felt that perhaps it was the absolute space or the fundamental frame of reference Newton was looking for and in which (or in a frame of reference fixed relative to it) his laws of motion would hold *perfectly.*

On the face of it, it was such an inviting —and exciting—proposition, and many a brilliant experiment were sought to be devised to establish the existence of this elusive medium, ether, and hence that of an absolute frame of reference,. It was but natural that light waves should be used for the purpose, since, after all, ether was little else but just a vehicle for them.

Now, the orbit of the earth a found the sun having a radius of 93 million miles, its orbital velocity works out to 18.5 mi or 30 km/sec, So that, even assuming that the sun has no ether drift, the velocity of the earth through the ether must be this orbital velocity of 18.5 mi or 30 kmfsec and the same must, therefore, also be the velocity of the resulting ether drift. Since the velocity of light is 186,000 mi or 3×10^5 km/sec, it will be easily seen that the velocity of the expected ether drift is just 18.5/186,000 (or $30/3 \times 10^{-4}$) = 10^{-4} of the velocity of light. Obviously, therefore, a very sensitive apparatus was called for, if the experiment was to succeed.

Maxwell had actually despaired of any terrestrial methods being sensitive enough for the purpose and started looking for any astronomical evidence that may come his way in support. But *A.A. Michelson*, of the USA, when he learnt about it in the year 1879, accepted the challenge and, along with his colleague *E.W. Morley* (at the Case *Institute of Technology* in Cleveland), succeeded in performing, in the year 1887, one of the most' famous experiments in the history of physics, (see next article) which gave a null result, striking at the very root of the ether hypothesis.

RESULTS FLOWING FROM LORENTZ TRANSFORMATION

Let us now examine some of the relativistic effects, *i.e.*, effects which are the result of relativistic as Lorentz transformation. Quite a few of them, as we shall see, will appear to be surprisingly new and strange because, in view of the small relative motion between the frames of reference of which we have experience in our daily life, we do not ordinarily come across any perceptible relativistic phenomena.

Let us consider a few important examples:

(i) *Length Contraction* : The length of a rod, as measured in a frame of reference, *at rest with respect to the observer*, is called its proper-length-and would obviously be the *same*. In all stationary frames of reference. It is only in moving reference frames that it will be different, depending upon their relative velocities with respect to the observer. Let us see how.

Consider a rod laid along, the axis of x in a frame of reference S, at rest with respect to the observer and let x_1 and x_2 be the coordinates of its two ends which may be noted at leisure, one after the other, the rod being at rest. Then, the *proper length* of the rod, say,

$$L_0 = (x_2 - x_1).$$

If the x-coordinates of the ends of the rod in the reference frame S', moving with a uniform velocity v with respect to frame S be x_1' and

x'_2, *as noted simultaneously at the same instant t'*, we have length of the rod in the moving reference frame S' given by $L = (x'_2 - x'_1)$.

To correlate L_0 and L, we note that inverse Lorentz transformation gives

$$x_1 = \frac{x'_1 + vt'}{\sqrt{1 - v^2/c^2}} \text{ and } x_2 \frac{x'_2 + vt'}{\sqrt{1 - v^2/c^2}}.$$

So that, $$L_o = x_2 - x_1 = \frac{x'_2 + vt'}{\sqrt{1 - v^2/c^2}} - \frac{\lambda'_1 + vt'}{\sqrt{1 - v^2/c^2}}$$

$$= \frac{x'_2 + x'_1}{\sqrt{1 - v^2/c^2}} = \frac{L}{\sqrt{1 - v^2/c^2}}$$

whence, $$L = L_o \sqrt{1 - v^2/c^2},$$

which is *clearly shorter than its proper length L_o*. And, as will be easily seen, the higher the value of v, *i.e.*, the faster the motion of the rod (or of the reference frame carrying it), with respect to the observer, the shorter its length compared with its proper length. So that, if $v = c$, *i.e., if the rod be moving with the velocity of light (c), its length would be zero.*

It follows at once, therefore, that the length of the rod or any other object when $v = 0$, *or when the rod or the object is at rest with respect to the observer, i.e., its proper length, is the largest.*

There will, of course, be no shortening of the length of the rod if it lies along the axis of y or z, perpendicular to the x-direction or the direction of motion of the frame of reference carrying it, though there will, naturally be some shortening if the length of the rod be inclined to the direction of its velocity for, then, it will have a component in the latter direction.

This shortening or contraction in the length of an object along its direction of motion is known as the Lorentz—Fitzgerald contraction, since both *Lorentz* and the Irish physicist *G.F. Fitzgerald*, during the 1890's, independently suggested that the null result of Michelson-Morley experiment could be accounted for, if the length of a material object contracted by the fraction $\sqrt{1 - v^2/c^2}$ in its direction of motion with speed v. through ether. Of course, as we know *now*, the contraction is certainly this fraction but for entirely different reasons.

It may be pointed out that there is *reciprocity of the contraction effect, i.e.*, the contraction works both ways—a length along the x-axis in frame S being contracted for the observer in S' and a length along the x'-axis in frame S' being contracted for the observer in S. For, if the rod be at rest along the axis of x' with respect to the moving frame S', its proper length, as measured by the observer in S' (at rest with respect to it) will be $L_o = (x'_2 - x'_1)$, where x'_1 and x'_2 are the coordinates of the end-points of the rod along this axis. And, if the x-coordinates of the two end-points of the rod in frame S, as noted *instantaneously, i.e.*, at the same instant t, by the observer in S', be x_1 and x_2, its length, as measured in frame S, is $L = (x_2 - x_1)$.

Since, in accordance with *Lorentz transformation*,

$$x'_1 = \frac{x_1 - vt}{\sqrt{1 - v^2/c^2}} \quad \text{and} \quad x'_2 \frac{x_2 - vt}{\sqrt{1 - v^2/c^2}}$$

we have $L_o = x'_2 - x'_1 = \dfrac{x_2 - vt}{\sqrt{1-v^2/c^2}} - \dfrac{x_1 - vt}{\sqrt{1-v^2/c^2}} = \dfrac{x_2 - x_1}{\sqrt{1-v^2/c^2}}$

$$= \frac{L}{\sqrt{1-v^2/c^2}}$$

whence, $L = L_o\sqrt{1-v^2/c^2}$,

Thus, *with relative motion between an object and t'v observer, the length of the object, as measured by the latter, always comes out to be shorter than its proper length.*

It is for this reason that if we have two frames of reference in relative motion (one stationary and the other in motion) along the x-direction, say, a straight line parallel to this direction in one appears shorter to an observer in the other,—not, so, however, a straight line along the y or the z-direction, perpendicular to the x-direction. Similarly, a square and a circle in one appear to the observer in the other to be a rectangle and an ellipse respectively Fig. 1.4, the sides of the square and the radius of the circle (or their components) in the direction of motion getting shortened.

As will be easily seen, however, for small values of v for which $v/c \to 0$, the factor $\sqrt{1 - v^2/c^2} = 1$, so that $L = L_o$, a result in accordance with Classical mechanics where length is treated as an absolute quantity, unaffected by rest or motion.

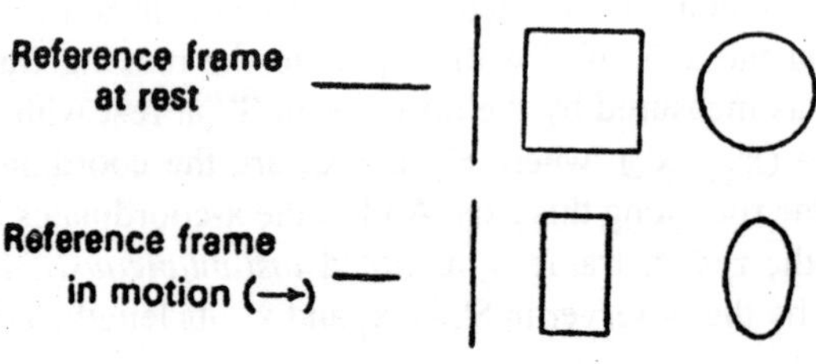

Fig. 1.4

This is exactly the reason why at speeds with which we are ordinarily concerned (and which are an infinitesimal fraction of c), the *Lorentz-Fitzgerald* contraction is nil or negligible.

For an interesting example (and evidence) of the contraction phenomenon, see worked example 14 (i) at the end of the chapter.

> *Note :* It is interesting to point out that even though a fast moving object must undergo *Lorentz-Fitzgerald* contraction, its length, if observed visually (*i.e.*, directly by the eye) or photographically, would *not* show any such contraction. Strangely enough, this phenomenon was discovered and studied only in the year 1959, more than half a century after the publication of the special theory of relativity. It may, however, be explained thus: The tight proceeding from the farthest part of the object leaves the earliest and that from its nearest part, the latest, with that from other parts at times in between these two extremes, to reach the eye or the camera *simultaneously* so as to form a composite image of the object on the retina of the eye or the film of the-camera. As can well be imagined, this should actually make the moving object look larger in size in is direction of motion It so happens, however, that this apparent increase in its length is just offset by the *Lorentz-Fitzgerald* contraction that it must undergo on account of its relative motion. As a, result, it appears to be neither lengthened nor shortened. Thus, if the moving object be a three-dimensional one, like a *cube*, no change in its shape will come about except a rotation of its orientation, which may approach 180° as $v \to c$. And, in the case of a sphere, there will be no distortion whatsoever.

(ii) *Time Dilation :* Time intervals too, like length, are affected by relative motion. Thus, to an observer, a clock appears to run slower when in motion, than when at rest with respect to him. So that, the time interval between two events occurring at a given point in space in a frame of reference, moving with respect to the observer, appears to be greater as noted on a clock at rest with respect to him than on an identical clock in the moving frame of reference itself. This is referred to as *time dilation,* meaning *apparent lengthening of lime.* Let us deduce this result from *Lorentz transformation.*

Suppose we have two frames of reference, S and S' Fig. 1.2, the former at rest and the latter moving with a uniform velocity v relative to it, along the +x direction. Suppose further that two identical clocks are carried by the two reference frames at their origins O and O', such that they both show zero time when just against each other, *i.e.*, when their origins just cross each other. Then, if two events occur at any given point x' in frame S', at times t_1' and t_2', *as noted on the clock carried by it* and at times t_1 and t_2 *as noted on the clock carried by frame S,* we clearly have *time interval between the two events, as noted on the clock in the moving frame S'* given by $\Delta r' = (t_2' - t_1')$.

And, time interval between the same two events as noted on the clock in the stationary frame S given by $\Delta t = (t_2 - t_1)$.

Now, in accordance with Lorentz transformation, we have

$$t_1 = \frac{t'_1 + vx'/c^2}{\sqrt{1 - v^2/c^2}} \quad \text{and} \quad t_2 = \frac{t'_2 + vx'/c^2}{\sqrt{1 - v^2/c^2}}$$

Therefore, $$\Delta t = (t_2 - t_1) = \frac{t'_2 + vx'/c^2}{\sqrt{1 - v^2/c^2}} - \frac{t'_2 + vx'/c^2}{\sqrt{1 - v^2/c^2}} = \frac{t'_2 - t'_1}{\sqrt{1 - v^2/c^2}}$$

$$= \frac{\Delta t'}{\sqrt{1 - v^2/c^2}} = \gamma \Delta t'.$$

Since the factor $1/\sqrt{1 - v^2/c^2}$ or y is greater than 1, we have $\Delta t > \Delta t'$.

Thus, *the time interval $\Delta t'$ between two events occurring at a given point in the moving frame S' appears to be longer, or dilated, by a factor* $1/\sqrt{1 - v^2/c^2}$ or γ to the observer in the stationary frame S.

The same will happen if frame S' were at rest and S in motion. So that, *like length contraction, time dilation too shows reciprocity effect, i.e.*, works both ways or *is independent of the direction of velocity and depends only on its magnitude*. In general, therefore, the interval between two events at the same point in a moving frame appears to be longer by a factor γ to an observer in a stationary frame.

This interval of time $\Delta t'$ between two events occurring at the same place, measured on the clock, moving with the reference frame S' in which the events occur, and hence at rest with respect to it, is called the *proper time interval* or the *proper time*, the usual symbol for which is τ. Thus, the proper time interval in a moving system is always less than the corresponding time interval, measured on a system at rest.

It follows from the above that the passage of time and hence also all physiological processes that go with it, like pulse and heart beats, and, intact, the process of ageing itself, are slowed down in a fast moving reference frame. Indeed, if the velocity of the moving frame be v = c, the passage of time and hence also the process of ageing will stop altogether!

This explains, in a rough and ready manner, what has come to be known as the *Twin paradox, viz.*, that, of two identical twins, if one goes about at a high speed in a rocket and the other stays behind on the earth, the former, when he returns back to the earth will be younger than the latter.

Incidentally, this also shows that nothing material can really travel faster than light,—a proven fact, having the force of a *fundamental law of nature*.

In case, however, the velocity (v) of the moving frame of reference be small, as is ordinarily the case, so that $v/c \rightarrow 0$, we have $\Delta t = \Delta t'$, *i.e.*, the time intervals between the two events in both the moving and the stationary systems are the same,—a result in accord with the absolute nature of time in Classical or Newtonian physics.

Most fundamental particles disintegrate spontaneously after a comparatively very brief spell of life, called their *life-time*. In a frame of reference S in which the particle is at rest (*i.e.*, its own frame of reference), the mean life-time is denoted by τ and is called the *proper* life-time of the particle. It is found to depend upon the velocity of the particle. Thus, in a frame of reference S' (say, the laboratory frame) in which it is moving with velocity v relative to S, the life-time gets dilated to

$$\tau' = \gamma\tau = \frac{\tau}{\sqrt{1 - v^2/c^2}}$$

Further, if N_o be the number of particles that undergo disintegration in the particle's own reference frame, the number of particles that survive up to time t is given by

$$N_{(t)} = N_o e^{t/\tau}$$

Verification of Time Dilation

(i) *Indirect Verification :* An indirect verification of the time dilation effect is afforded by the fact of μ = *mesons* or *muons*, for short, produced higher up in the atmosphere, reaching the ground level in the very short span of their life-time.

These elementary particles, having a mass 215 times that of an electron, are produced in the upper reaches of the atmosphere as a result of collision of cosmic rays with the air molecules and though they do not all have the same velocity, the faster ones among them have a velocity 0.998c, very close to that of light in free space. Even, with this high velocity, however, they can, in their whole life-time of 2.2×10^{-6} sec cover at the most a distance equal to $0.998c \times 2.2 \times 10^{-6} = 658.68$ *metres* before they disintegrate or decay into an *electron* and two *neutrinos* each (a neutrino being a particle of zero charge and zero mass). And, yet they are found copiously in the laboratory 10 km below the spot where they are produced.

How is this possible? Obviously, because of time dilation. For, whereas in their own frame of reference (*i.e.*, the frame of reference in which they are produced), their mean life-time T is only 2.2×10^{-6} sec, for us, in the laboratory frame of reference, it is dilated to

$$\tau' = \gamma\tau = \frac{\tau}{\sqrt{1 - v^2/c^2}} = \frac{2.2 \times 10^{-6}}{\sqrt{\frac{1 - (0.998c)^2}{c^2}}} = 34.38 \times 10^{-6} \text{ sec},$$

than 15 times that in their own frame of reference. In this much longer life-time, they can easily cover a distance of $0.998c \times 34.38 \times 10^{-6} = 10.29$ few. This at once explains their presence in the laboratory, 10 km below, at the ground level.

(ii) *Direct Verification :* Directly, the time dilation effect may be verified in the laboratory, using a beam of π^+ *mesons or Pions*, for short, (the products of highly energetic nuclear reactions in the upper atmosphere) which are produced artificially in the laboratory by bombarding a suitable target of *carbon* or *beryllium* by highly energetic protons or α-particles from a *synchrocyclotron.* These particles are comparatively heavier than mesons, with a mass 273 times that of an electron, but have a much shorter life-time τ of 2.6×10^{-8} sec is their own frame of reference, after which they disintegrate into a *meson* and two *neutrinos* each. Though all the $\pi^\pm$ mesons do not have the same velocity, the faster ones may be assumed to have a velocity of 0.9c. So that, their dilated life-time τ in the laboratory frame of reference should be equal to

$$\gamma\tau = \frac{2.6 \times 10^{-8}}{\sqrt{1 - v^2/c^2}} = \frac{2.6 \times 10^{-8}}{\sqrt{1 - \frac{(0.9c)^2}{c^2}}} = 5.96 \times 10^{-8} \text{ sec, i.e.,}$$

more than twice that in their own frame of reference (in which they are at rest).

To verify whether this is so, we obtain a beam of *pions* in the manner indicated (*i.e.*, by bombarding a ribbon target T of *carbon* or *beryllium* Fig. 1.5 by α-*particles* or *protons* from a synchrocyclotron). This beam is narrowed down by passing it through a *collimator slit* S, as shown. The pions in the beam decay into *muous* and *nutrinos* and the muons thus produced are detected by a suitable device D, alongside which the beam passes.

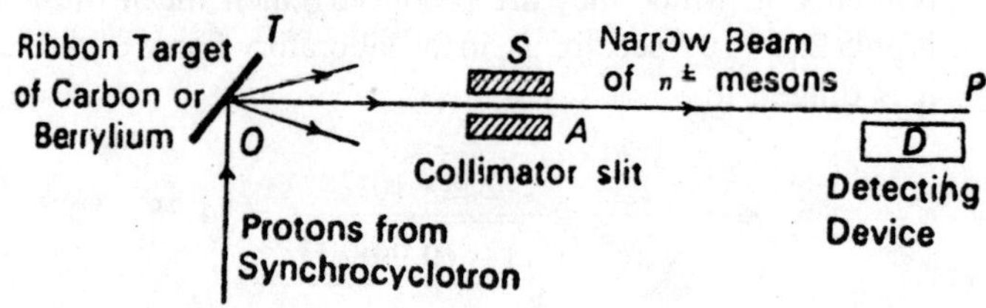

Fig. 1.5

Obviously, the number of *muons* recorded by the detecting device per second is proportional to the number of pions present in the beam in the vicinity of the device, *i.e.*, to the flux of pions adjacent to it. By observing the recordings of the detecting device at various points along the path OP of the beam, therefore, the dilated life-time τ' of the moving pions can be obtained.

To obtain the mean life-time or the *proper life-time* τ of the pions (*i.e.*, their life-time in their own frame of reference, in which they are at rest), they are brought to rest at A right at' the beginning of their path as they emerge from the slit, by introducing a thick solid slab of a suitable material. The time interval elapsing between the entering of the pions in the solid slab and the emerging of muons therefrom, noted electronically, then gives the proper life-time τ of the pions constituting the beam.

It is found that $\tau' = \gamma\tau = \dfrac{t}{\sqrt{1 - v^2/c^2}}$ and the time dilation effect thus stands verified.

(iii) *Simultaneity* : By *simultaneity* of two events we understand their occurrence at *exactly the same time*. Thus, for example, if in a fixed frame of reference S, A and B are two points distant x_1 and x_2 respectively from the origin 0 and P, the carefully determined mid-*point* of AB. Then, if a light signal is emitted at P, it-will arrive at A at exactly the same time as at B, since the velocity of light (c) is the same in all directions. We, therefore, say that the two events, viz., the arrival of the light signal at A and at B, occur simultaneously at time t, say, *i.e.*, one event occurs at x_1 at time t and the other at x_2 also at the same time r.

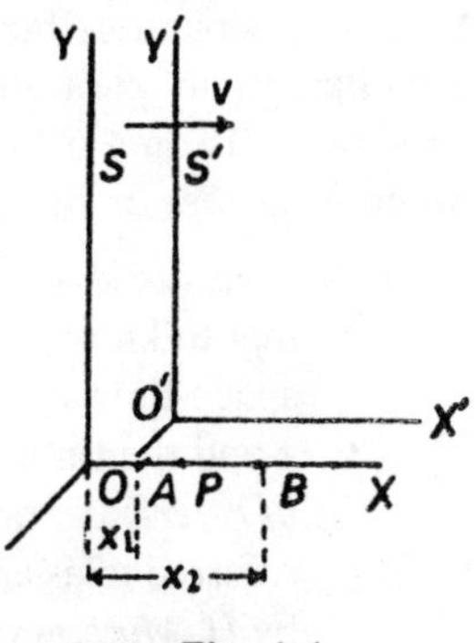

Fig. 1.6

Let us see if the two events also appear to be simultaneous to an observer in a frame of reference S', moving relative to S, with velocity v along the positive direction of its x-axis. For the observer in S, the corresponding values of t for the events at x_1 and x_2 will, in accordance with Lorentz transformation, be given respectively by

$$t'_1 = \gamma\left(t - \frac{vx_1}{c^2}\right) \quad \text{and} \quad t'_2 = \gamma\left(t - \frac{vx_2}{c^2}\right)$$

So that, the time interval between the two events, as observed by the observer in S' wilt be $t' = (t'_1 - t'_1) = \gamma^v(x_1 - x_2)/c^2$, which is surely not zero, indicating that the two events at A and B, which appear to be simultaneous to the observer in *S do not appear so* to the observer in S' but to have a time interval t' between them, the magnitude of which depends upon both, the distance between the two events and the velocity of frame S' relative to S.

Exactly in the same manner it can be shown that two events that appear to be simultaneous to an observer in frame S' do not appear to be so to the observer in frame S.

As can welt be seen, *this failure of simultaneity in one frame of reference in relative motion with another is a direct consequence of the value of c being the same in all frames of reference and in all directions.* So that, if we accept, as we must, the postulate of *invariance* of the *velocity of light*, we have also to accept the consequence-flowing from it, namely, the relativity of simultaneity, *i.e.*, the simultaneity of events, separated in space, not being something absolute that can be expected or accepted in all frames of reference. It is there in each frame of reference separately but there can be no question of agreement between simultaneity in one frame with that in another in relative motion with it. In short, it is only a *local affair* of each individual frame of reference, whether at rest or in motion.

(iv) *Invariance of Space-Time Interval* : It has already been mentioned that to know the true position of a point, we must know its four space-time coordinates, viz., the three spatial coordinates x, y, z and the time coordinate t. Such a point in a *space-time frame of reference*, or an occurrence there, is, as we have seen before, referred to as an *event* in the space-time language, first proposed by *H. Minkowski* in the year 1908. And, just as the familiar space vectors enable us to deduce spatial relationships between points or lines, independently of any particular coordinate system, so also, with the addition of a time vector to them, the four space-time vectors enable us to express physical laws in special relativity (involving terms measured in space-time coordinates), irrespective of any particular frame of reference, *i.e., laws that are invariant to Lorentz transformation.*

Now, before the advent of the theory of relativity, as we know *space* and *time* were regarded as *separate entities* and both space as well as time intervals were invariant to Galilian transformation, their values being thus quite independent of the frame of reference used, This is not the case with Lorentz transformation, under which neither of them is invariant. What remains invariant under this transformation is the combination $c^2t^2 - x^2 - y^2 - z^2$, as we have seen. It is called the *space-time* interval between two points, one of which has been taken at the origin itself. It is usually denoted by the symbol $S^2{}_{12}$ and, as is so obvious, it is actually the *interval squared*, though, commonly, we refer to it as the interval.

In case one of the points is not at the origin, and the coordinates of the two points are x_1, y_1, z_1, t_1 and x_2, y_2, z_2, t_2 respectively, the *space-time* interval is obviously $c^2\left(t_2^2 - t_1^2\right) - (x_2 - x_1)^2 - (y_2 - y_1)^2 - (z_2 - z_1)^2$. That this is invariant to Lorentz transformation may be easily seen. For, if the corresponding coordinates of these points in a frame S', moving along the axis of x of frame S with a constant velocity v relative to it, be x'_1, y'_1, z'_1, t'_1 and x'_2, y'_2, z'_2, t'_2, ts we have, in accordance with Inverse Lorentz transformation,

$$S'^{1}_{12} = c^2\left(t_2^2 - t_1^2\right) = c^2\gamma^2\left(t'_2 - t'_1\right) + v\left(x'_2 - x'_1\right)/c^2]^2$$

$$(x_2 - x_1) = \gamma^2\left[\left(x'_2 - x'_1\right) + v\left(t'_1 - t'_1\right)\right]^2, (y_2 - y_1)^2 = \left(y'_2 - y'_1\right)^2$$

and $$(z_2 - z_1)^2 = \left(z'_2 - z'_1\right)^2$$

$$\therefore \quad c^2 = \left(t'_2 - t'_1\right) - (x_2 - x_1)^2 - (y_2 - y_1)^2 - (z_2 - z_1)^2$$

$$= c^2\gamma^2\left[\left(t'_2 - t'_1\right) + v\left(x'_2 - x'_1\right)/c^2\right]^2 - \gamma^2\left[\left(x'_2 - x'_1\right) + v\left(t'_2 - t'_1\right)\right]^2 - \left(y'_2 - y'_1\right)^2 - \left(z'_2 - z'_1\right)^2$$

Or, $$S'^2_{12} = c^2\left(t'_2 - t'_1\right)^2\gamma^2\left(1 - \frac{v^2}{c^2}\right) - \left(x'_2 - x'_1\right)^2\gamma^2\left(1 - \frac{v^2}{c^2}\right) - \left(y'_2 - y'_1\right)^2 - \left(z'_2 - z'_1\right)^2$$

Since $$\gamma = 1/\sqrt{1 - v^2/c^2},$$

Or, $g^2 = 1/(1 = v^2/c^2)$

we have $\gamma^2\left(1-\frac{v^2}{c^2}\right)=1$

$$\therefore \quad S_{12}^{'2} = c^2\left(t_2^{'}-t_1^{'}\right)^2-\left(x_2^{'}-x_1^{'}\right)^2-\left(y_2^{'}-y_1^{'}\right)^2-\left(z_2^{'}-z_1^{'}\right)^2$$

Thus,

$$c^2\left(t_2-t_1\right)^2-\left(x_2-x_2\right)^2-\left(y_2-y_2\right)^2-\left(z_2-z_2\right)^2 c^2=\left(t_2^{'}-t_1^{'}\right)^2 -\left(x_2^{'}-x_1^{'}\right)^2-\left(y_2^{'}-y_1^{'}\right)^2-\left(z_2^{'}-z_1^{'}\right)^2$$

Or, $$S_{12}^2 = S_{12}^{'2}, \qquad ...(i)$$

clearly showing that *the space-time interval is invariant to Lorentz transformation.*

It may be noted again that the *space and the time-intervals, taken separately are not invariant.* For, clearly, $(x_2 - x_1)^2 + (y_2 - y_1)^2 + (z_2 - z_1)^2$ is not equal to $\left(x_2^{'}-x_1^{'}\right)^2+\left(y_2^{'}-y_1^{'}\right)^2+\left(z_2^{'}-z_1^{'}\right)^2$ and nor is $(t_2 - t_1) = (t_2' - t_1')$.

Thus, whereas in Classical physics, *time* and *space* were taken to be two different things, the Lorentz transformation equations and relation (i) above, proclaim clearly that *in the theory of relativity, time* and *space are equivalent* and can be expressed one in terms of the other. Thus, it can be seen from equation (i) that, *in terms of distance*, 1 sec is 3×10^8 *metre*, the distance covered by light in free space in 1 sec and, similarly, 1 *metre of time* is $1/3 \times 10^8$ sec, the time taken by light to cover a distance of one metre.

Thus, measuring time and space in the same units in a system in which c = 1, the Lorentz transformation equations, as also equation (i) above are greatly simplified and respectively become

$$x' = (x - vt)/\sqrt{1-v^2},\ y' = y,\ z' = z$$

and $$t' = (t - vx)/\sqrt{1-v^2}, \qquad ...(ii)$$

and $$(t_2 - t_1)^2 - (x_2 - x_1)^2 - (y_2 - y_1)^2 - (z_2 - z_1)^2$$
$$= (t_2' - t_1')^2 - (x_2' - x_1')^2 - (y_2' - y_1')^2 - (z_2' - z_1')^2 \qquad ...(iii)$$

Now, if we consider two points (or particles) which, in a given co-ordinate system, have only space but no time (or zero time), we can see from relation (iii) that the interval squared would be negative and the interval, therefore, an *imaginary* one. Such an interval is referred to as a *space-like interval*, obviously because the interval, in this case, is more like space than like time. On the other hand, when we have two points, or particles, occupying the same place in the coordinate system but having different times, the square of the time being positive and the distance zero, the interval squared is positive and hence the interval real. Such an interval is called a *time-like interval.*

Incidentally, the invariance of the space-time interval in two frames of reference means, in other words, the invariance of the value of c in the two frames.

(v) *Transformation of Velocity—Velocity Addition* : Suppose a particle has a velocity u in a frame of reference S at rest, such that its components along the three coordinate axes are $u_z = d^x/dt$, uy = dy/dt and $u_z = dz/dt$ and its velocity in a frame of reference S', moving with a uniform velocity v relative to S along the positive direction of the x-axis, is u', with its components $u_x' = dx'/dt$, $u_y' = d_y'/dt$, and $u_z' = dz'/dt$ along the three coordinate axes, as measured in S'. Let us see how these component velocities in the two frames of reference are related to each other.

As we know, in accordance with Lorentz transformation,

$$x' = \gamma(x - vt),\ y' = y,\ z' = z$$

and

$$t' = \gamma(t - vx/c^2),$$

where

$$\gamma = \sqrt{1 - v^2}.$$

So that,

$$u_x' = \frac{dx'}{dt} = \frac{\gamma(dx - vdt)}{\gamma(dt - vdx/c^2)} = \frac{dx - dvt}{dt - vdx/c^2}$$

Dividing the numerator and the denominator on the right hand side of the equation by dt, we have

$$u_x' = \frac{dx'}{dt} = \frac{dx/dt - v}{1 - \frac{v}{c^2} \cdot \frac{dx}{dt}}$$

Or, substituting u_x for dx/dt, we have $u_x' = \dfrac{u_x - v}{1 - vu_x/c^2}$

And, $$u_y' = \frac{dy'}{dt'} = \frac{dy}{\gamma(dt - vdx/c^2)}.$$

Or, dividing the numerator and denominator on the right hand side of the equation by dt, as before, we have

$$u_y' = \frac{dy/dt}{\gamma\left(1 - \frac{v}{c^2}\cdot\frac{dx}{dt}\right)} = \frac{u_y}{\gamma(1 - vu_x/c^2)} \qquad \left[\because dy/dt = u_y \text{ and } dx/dt = u_x\right]$$

Similarly, $$u_z' = \frac{u_z}{\gamma(1 - vu_x/c^2)}$$

where u_x', u_y' and u_z' are the components of the resultant velocity u in frame S'.

In case velocity u is along the axis of x, its components u_y and u_z in frame S and hence also u_y' and u_z' in frame S' are each equal to zero and we have

$$u' = \frac{u - v}{1 - uv/c^2}$$

It will thus be seen that, as under Galilian transformation, so also here, *velocity is not invariant in the two frames reference.*

By *inverse Lorentz transformation* (*i.e.*, replacing v by – v), we obtain

$$u_x = \frac{u_x' + v}{1 - vu_x'/c^2}, \quad u_y = \frac{u_y'}{\gamma(1 - vu_x'/c^2)}$$

and $$u_z = \frac{u_z'}{\gamma(1 - vu_x'/c^2)}$$

Again, if the velocity of the particle, u', be directed along the x'-axis in frame S', its components u_y' and u_z' in frame S' and hence also u_y and u, in frame S are each equal to zero and we have

$$u = \frac{u' + v}{1 + uv/c^2}$$

N.B. : Quite often it is more convenient to measure and express high velocities as fractions of c rather than in cm or m/sec. So that, we divide the relations for u_x' and u_x say, by c and put these in the form

$$\frac{u'_x}{c} = \frac{(u_x/c) - (v/c)}{1 - \frac{v}{c}.\frac{u_x}{c}}$$

and

$$\frac{u_x}{c} = \frac{(u'_x/c) + (v/c)}{1 - \frac{v}{c}.\frac{u'_x}{c}}$$

Other relations for u'_y, u'_z, u'_y and u_z may also be similarly expressed.

Now, the velocity u of the particle in frame S may be regarded as the resultant of its velocity u in frame S' and the velocity v of frame S' relative to S. So that, the expressions above give us a method for the addition of high or relativistic velocities (*i.e.*, velocities that approach the value of c) and constitute a *law for the addition of relativistic velocities*. Thus, from relation (B) above, we may formally state the law as follows:

If a particle moves with velocity u' in a reference frame S' and if S' has a velocity v relative to a frame S, the velocity of the particle relative to S is given by $(u' + v)/(1 + u'v/c^2)$.

It may be noted that

(i) If u'_x be very much less than c, so that $u'_x/c \rightarrow 0$, the above expressions all reduce to those obtained by Galilian transformation, i.e., $u_x = u_x' + u_y$, $u_y' = u_y'$ and $u_z = u'_z$.

(ii) For higher values of u_x', the denominator in the expression for u_x becomes greater than 1 and hence the resultant velocity u_x is smaller than $\left(u'_x + v\right)$, the value given by Galilian transformation.

(iii) If $u'_x = c$, i.e., if the velocity of the particle along the x'-axis in frame S' be equal to that of light (i.e., if the particle be a photon), we have $u_x \dfrac{c + y}{1 + yc/c^2} = \dfrac{c + v}{(c + v)/c} = c$, i.e., the particle has the same velocity c in frame S.

Thus, *this particular velocity c alone is invariant in the two frames of reference irrespective of their relative velocity v.*

But this result is only to be expected in view of the fact that *Lorentz transformation* itself has been obtained on the basic assumption of the constancy of the value of c in all reference frames, whatever their own velocities.

Incidentally, another interesting result that emerges is that, *under Lorentz transformation, no two velocities can add up to more than the value of c.* The same is not true under Galilian transformation. The following example will illustrate the point.

Suppose the relative velocity of S' with respect to S, along its x-axis, *i.e.*, $v = 3c/4$ and a particle in S moves along the +x'-direction with velocity c/2, *i.e.*, $u_x' = c/2$. Then, in accordance with Galilian transformation, the resultant velocity of the particle, relative to frame S will be $u_x = u_x = \frac{c}{2} + \frac{3c}{4} = \frac{5c}{4} = 1.25\,c$, *i.e.*, greater than c.

But, under *Lorentz transformation*, it will be

$$\frac{u_x}{c} = \frac{(u_x'/c) + (v/c)}{1 + \frac{v}{c} \cdot \frac{u_x'}{c}} = \frac{\left(\frac{1}{2}\right) + \left(\frac{3}{4}\right)}{1 + \frac{3}{4} \cdot \frac{1}{2}}$$

$$= \frac{5}{4} \times \frac{8}{11} = \frac{10}{11},$$

whence, $u_x = \frac{10}{11} c$, *i.e., less than c.*

(vi) *Relativistic Doppler Effect* : We are already familiar with the *Doppler effect* in Sound, viz., the apparent change in the frequency of sound due to motion of the source of sound and the listener relative to the medium. It bears the name of *Doppler* (an Austrian) who was the first to have worked out the theory in the case of sound waves in the year 1842. He also called attention to the relevance of the phenomenon in the case of light.

The Doppler effect in light is, however, different from that in sound for the simple reason that, unlike sound, light requires no material medium for its propagation and hence its relative velocity is the same (c) for all observers regardless of their own state of motion. Then, again, the velocity of light is so great that any apparent change in the frequency or the wavelength of a light pulse is appreciable only at relativistic needs, *i.e.*, speeds near about c. Let us then proceed to discuss the case of light.

Imagine two reference frames S and S', where S is stationary and S' in relative motion with respect to S, with a constant velocity v along the axis of x, (Fig. 2.7). Let two light signals or pulses be emitted from a source placed at the origin O in frame S at time t = 0 and t = T, where T is the *true time period* of the light pulses. Suppose the interval between the reception of these pulses by an observer at the origin O' in frame S' is Δt'. Then, clearly, Δt' is the *proper lime interval* T' between the two pulses, as measured in frame S' (since they are both observed at the same point O' by an observer at rest with respect to S'). The observer will naturally interpret this interval as the time-period of the pulses received by him, *i.e.*, the apparent time-period of the pulses for the observer in frame S' will be T'.

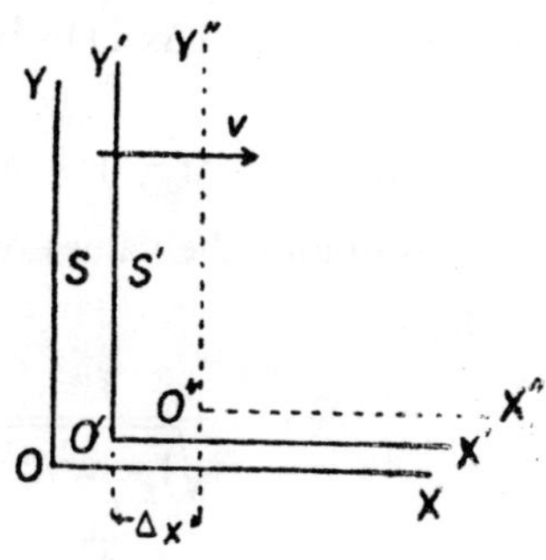

Fig. 1.7

Since the observer continues to be at O' all the time, the distance Δx' covered by him *in frame* S' during the reception of the two pulses is zero. Let us obtain the values Δx and Δt in frame S which correspond to those of Δx' and Δt' in frame S'.

From Inverse Lorentz transformation equation $x = \dfrac{x' + vt'}{\sqrt{1 - v^2/c^2}}$,

we have

$$\Delta x = \frac{\Delta x' + v\Delta t'}{\sqrt{1 - v^2/c^2}},$$ because the time-intervel here is Δt'.

Since Δx' = 0, we have $\Delta x = \dfrac{v\Delta t'}{\sqrt{1 - v^2/c^2}} = \dfrac{vT'}{\sqrt{1 - v^2/c^2}}$

$[\because \Delta t' = T'$

Clearly, the second pulse has to cover this much distance more than the first pulse in frame S, along, the axis of x, to be able to reach the observer at the origin O' in the moving frame S'.

Similarly, from the Inverse Lorentz transformation equation

$$t = \frac{t' + v_x'/c^2}{\sqrt{1 - v^2/c^2}},$$

we have $$\Delta t = \frac{\Delta t' + v\Delta x'/c^2}{\sqrt{1 - v^2/c^2}}$$

Or, since $\Delta x' = 0$, $\Delta t = -\dfrac{\Delta t'}{\sqrt{1 - v^2/c^2}} = \dfrac{T}{\sqrt{1 - v^2/c^2}}$

This obviously includes both the actual time-period T of the pulses and the time taken ($\Delta x/c$) by the second pulse to cover the extra distance Δx in frame S.

So that $\Delta t = T + \Delta x/c$.

Substituting the values of A/ and Ax obtained above, we therefore have

$$\frac{T'}{\sqrt{1 - v^2/c^2}} = T + \frac{vT'}{\sqrt{1 - v^2/c^2}}$$

whence, $$\frac{T'}{\sqrt{1 - v^2/c^2}} = T + \frac{vT'}{\sqrt{1 - v^2/c^2}}$$

If v and v' be the *actual* and the *observed* (or *apparent*) frequencies of the light pulses respectively (*i.e.*, their frequencies in frames S and S'), we have $v = 1/T$ and $v' = 1/T'$. So that,

$$\frac{1}{v} = \frac{(1 - v/c)}{\sqrt{1 - v^2/c^2}}.$$

Or, $$v' = v\,\frac{(1 - v/c)}{\sqrt{1 - v^2/c^2}}$$

$$= v\,\frac{(1 - v/c)}{\sqrt{(1 - v/c)\,(1 + v/c)}}$$

Or, $$v' = v\sqrt{\frac{1 - v/c}{1 + v/c}} = v\sqrt{\frac{1 - \beta}{1 + \beta}} \qquad [\because v/c = \beta.$$

This is known as *Doppler's formula* and gives the apparent frequency v' of the light pulses of actual frequency v as observed in frame S'.

Since the velocity of light $c = v\lambda = v'\lambda'$, we have $v = c/\lambda$ and $v' = c/\lambda'$, where λ and λ' are the wavelengths of the light pulses corresponding to frequencies v and v' respectively. From relation I, therefore, we have

$$\frac{c}{\lambda'} = \frac{c}{\lambda}\sqrt{\frac{1 - v/c}{1 + v/c}} = \frac{c}{\lambda}\sqrt{\frac{1 - \beta}{1 + \beta}}$$

Or, $$\lambda' = \lambda\sqrt{\frac{1 - v/c}{1 + v/c}} = \lambda\sqrt{\frac{1-\beta}{1+\beta}},$$

which gives the apparent wavelength λ' of the light pulses, of actual wavelength λ, as observed in frame S'.

In both relations I and II, v is positive or negative according as the observer is receding from, or approaching, the source of light; so that, in the former case, the apparent frequency of the pulse (ν') decreases and its apparent wavelength (λ') increases and in the latter case, the exact opposite happens, i.e., the apparent frequency increases and the apparent wavelength decreases.

N.B. : It will be readily seen that if we ignore the second and higher order terms in v and c, we obtain, from I above, the relation $\nu' = \nu(1 - v/c)$ for the apparent frequency. This is a case of *non-relativistic Doppler effect*, the same as in the case of sound.

Now, what we have discussed above is, in point of fact, the longitudinal Doppler effect. There is also a *transverse Doppler effect* in the case of light, though it has no non-relativistic counterpart in the case of sound. This effect relates to the frequency observed in a direction at right angles to that of the motion of the source, usually an *atom*. The apparent frequency is here given by $\nu' = \nu\sqrt{1 - v^2/c^2}$, where ν is the actual frequency of the light pulse emitted by an atom in the frame of reference in which it is at rest and ν', the apparent frequency as noted in a frame of reference moving relative to it with velocity v.

Experimental confirmation of Doppler effect : The fact that there is indeed an apparent change in the frequency or the wavelength of light due to Doppler effect has been fully confirmed experimentally by *Ives* and *stilwell* in the year 1941.

They measured spectroscopically the change or shift in the average wavelength of a particular spectral line emitted by the hydrogen atom. This they succeeded in doing by using a beam of hydrogen atoms in a highly excited state. With the help of a very strong electric field, they accelerated the atoms as molecular (H_2^+) ions and as H_3^+ ions to a velocity of about 0.005c and then examined the spectrum of the atomic hydrogen, produced as a break-up product of these ions.

One particular line in the spectrum was closely studied and its wavelength obtained both in the forward and backward direction, *i.e.*, in the direction of motion of the atoms as well as in the opposite direction —the former by receiving the spectrum directly on the slit of the spectrometer and the latter, by reflecting the spectrum on to the slit with the help of a plane mirror. Also, the wavelength of the same spectral line was obtained for such of the hydrogen atoms as were at rest, of which there will always be quite a few.

If the wavelength of the spectral line in question, for the hydrogen atoms at rest, be λ_o and its wavelengths in the forward and backward directions be respectively λ_F and λ_B, we have.

average (or mean) apparent wavelength of the displaced line,

$$\lambda_{av} = (\lambda_F + \lambda_B)/2.$$

So that, *apparent change in the average wavelength of the line*

$$= \lambda_{av} - \lambda_o = 1/2\ (\lambda_F + \lambda_B) - \lambda_o.$$

But, as explained above, $\lambda_F = \lambda_o \sqrt{\dfrac{1-\beta}{1+\beta}}$ and $\lambda_B = \lambda_o \sqrt{\dfrac{1+\beta}{1-\beta}}$

$$\therefore\quad \lambda_{av} - \lambda_o = \frac{1}{2}\lambda_o\left(\sqrt{\frac{1-\beta}{1+\beta}} + \sqrt{\frac{1+\beta}{1-\beta}}\right) - \lambda_o = \frac{\lambda_o}{(1-\beta^2)^{1/2}}$$

$$= \lambda_o\ (1-\beta^2)^{-1/2} - \lambda_o$$

$$= \lambda_o(1 + 1/2\ \beta^2 + ...) - \lambda o = \frac{\lambda_o \beta^2}{2},\ \text{neglecting higher terms of } \beta.$$

Or, $$\lambda_{ov} - \lambda_o = \frac{\lambda_o v^2}{2c^2}.$$

Since the shift in wavelength $(\lambda_{ov} - \lambda_o)$, here, depends upon v^2, it is quite independent of the sign of v, *i.e.*, is in the same direction, whether v is positive or negative.

As calculated by this relation, *Ives* and *Stilwell* obtained the value 0.072 *Angstrom unit* or 0.072×10^{8} cm, whereas their observed value was found to be 0.074 *Angstrom unit* or 0.074×10^8 cm.

The excellent agreement between the two values thus confirms the validity of relativistic Doppler effect, as expected.

The recessional red shift : Every element shows its own characteristic lines of definite wavelengths in its spectrum. The spectral lines of all known elements have, therefore, been carefully mapped out to enable us to see at a glance the position and wavelength of the spectral line or lines emitted by any given element.

Now, in the spectrum of the light received from distant stars and galaxies, the characteristic spectral lines of the various known elements present in them are all found to be shifted by various amounts from their normal positions towards the lower frequency or the high wavelength side, namely, the red end of the spectrum.

This shift is attributable to *Doppler effect*, according to which $\lambda' = \lambda\sqrt{(1+\beta)/(1-\beta)} = \lambda\sqrt{(1+v/c)/(1-v/c)}$; so that, for a positive value of velocity v of the source (i.e., with the source receding from the observer), there is an apparent increase, and for a negative value of v (*i.e.*, with the source approaching the observer) there is an apparent decrease, in the wavelength of light received by the observer. In the former case, therefore, there is^ an apparent shift of the wavelength towards the red end, and in the latter case, towards the violet end of the spectrum. In fact, the velocity of recession or approach of the source may be easily determined from this shift in the wavelength of a spectral line.

The fact that light from the-distant stars and galaxies shows this shift of the spectral lines towards the red means that they are receding from us. Thus, since the shift of the spectral lines towards the red is due to the recession of the stars and galaxies from us on the earth, the phenomenon is called *recessional red shift* and lends strong support to the theory that the universe is expanding, with all bodies receding from one another. The Doppler effect has thus been instrumental in the detection, for the first time, of the apparent expansion of the universe.

On the basis of a large number of observations made on several galaxies, it is surmised that the relative velocity of a galaxy, at a distance r from us, is given by

$$V = \alpha r,$$

where a is an *empirical constant*, whose value is found to be .near about 3×10^{-18} sec^{-1}.

Interestingly enough, the reciprocal of α, *i.e.*, 1/α, which has the dimensions of time and which is equal to $1/3 \times 10^{-18} = 3 \times 10^{17}$ sec $= 10^{10}$ years, is spoken of as the *age of the universe* and its product with c, *i.e.*, c/α, which has the dimensions of length or distance and which is equal to $3 \times 10^{17} \times 3 \times 10^{10}$ or 10^{28} years, is similarly spoken of as the *radius of the universe*, though no one yet seems to know for certain what the significance of these terms actually is.

(vii) *The Relativity of Mass—Conservation of Momentum—Force* : The mass of a body too, like length (or space) and time, is dependent on the motion of the body. Its variation with velocity is, in fact, a direct consequence of the dilation of time, as will be clear from the following.

We know that for all inertial frames, in relative motion along the axis of x, the displacements Δy and Δz of a particle along the axes of y and z respectively remain unaffected whatever the value of the relative velocity v of the moving reference frame with respect to the stationary one.

The time taken by the particle to traverse this displacement Δy, and hence also its velocity component along the y-axis will, however, depend upon the frame of reference in question (since neither time nor velocity is invariant to Lorentz transformation).

But, again, the *proper time* $\Delta\tau$ (as noted on a clock carried by the moving frame itself) to cover the displacement Ay will, as we know, be the same in either reference frame, and so will, therefore, be the quantity $\Delta y/\Delta\tau$.

So that, if we take $\Delta y/\Delta\tau$ to be the velocity component of the particle along the axis of y in the moving frame and m_0 its rest mass or *proper mass, i.e.*, its mass as taken by an observer at rest with respect to it (*i.e.*, by the observer in the moving frame, in this case), we have *y-component of the momentum of the particle in the moving frame, i.e.*,

$$p_y = m_0 \Delta y/d\tau.$$

Now, if Δt be the time taken to cover the displacement Δy, as noted on an identical clock by an observer in the stationary frame, we have by Lorentz transformation, $\Delta\tau = Dt\sqrt{1 - v^2/c^2}$ whence,

$$\Delta t = \Delta\tau\sqrt{1 - v^2/c^2}$$

Thus, y-component of the velocity of the particle, as measured in the stationary frame, is $v_y = \Delta y/\Delta t$ and, therefore,

y-component of the momentum of the particle in the stationary frame = mv_y, *where m is the mass of the moving particle as taken by the observer in the stationary frame.*

Since, despite relativistic effects, the basic laws of physics in all frames of reference must have the same mathematical forms, *the law of conservation of momentum must hold good in either frame.* We, therefore, have

$$p_y = mv_y = m_o \frac{\Delta y}{\Delta \tau} = \frac{m_o \Delta y}{dt} \frac{dt}{d\tau} = m_o \frac{\Delta y dt}{dt\, d\tau} = m_o v_y \frac{dt}{d\tau},$$

whence, $m = m_o dt/d\tau$.

or, substituting the value of $dt/d\tau$ from above, we have

$$m = \frac{m_o}{\sqrt{1 - v^2/c^2}}$$

Since $\sqrt{1 - v^2/c^2}$ is always greater than 1, $m > m_o$,

i.e., the mass of a particle in a moving frame, as taken by an observer in a stationary frame, is always greater than its rest mass.

Here too we have the *reciprocity effect*; so that, for an observer moving with the particle, its mass will be w. but the mass of an identical particle in the stationary frame will appear to be

$$\sqrt{1 - v^2/c^2}$$

Alternatively, we may arrive at the same result as follows:

Suppose we have a reference frame S at rest (Fig. 1.8), in which two identical particles A and B, (two *protons*, say) are made to under-go a *symmetrical, glancing collision*, moving with equal and opposite velocities, as shown in Fig. 1.8(a), such that a line (shown dotted) parallel to the x axis bisects the angle between their trajectories. If the collision be a perfectly elastic one the x-components of the velocities of the two particles remain the same respectively, in magnitude and direction, both before and after the collision and since the particles have equal masses, *there is no change of momentum* in the *x-direction.* Again, because their components, after collision are simply reversed in direction, their magnitude remaining unchanged, the change in momentum of particle A is equal and opposite to the change in momentum of particle B (their initial y-components being oppositely directed), so that *the change in momentum along the y-direction too is zero* and *the momentum of the system thus remains conserved* in accordance with Newtonian physics, under which the masses of the particles remain unaffected by their velocities.

Imagine now that we observe this collision from a frame of reference S', moving relative to the stationary frame Σ with a velocity *equal to the horizontal or the x-component of particle B;* so that B has now a vertical velocity $u_y B$, say, its horizontal or x-component having been reduced to zero. It, therefore, appears to move up with velocity $u_y B$ and bounce back after collision with particle A, with its velocity reversed, as shown in Fig. 1.8(b). Particle A, however, has a horizontal component of velocity v, say, and a vertical component $v_y A$, the former remaining unchanged in magnitude and direction and the latter getting reversed in direction after collision.

Similarly, if we observe the collision from a frame S' moving with velocity v (equal to that of particle A in frame S), relative to frame S, we find that, instead of particle B, particle A now loses its horizontal or x-component of velocity and merely moves down with velocity $-u'_y A$, say, and bounces back after collision with particle B, with its velocity reversed, as shown in Fig. 1.8(c).

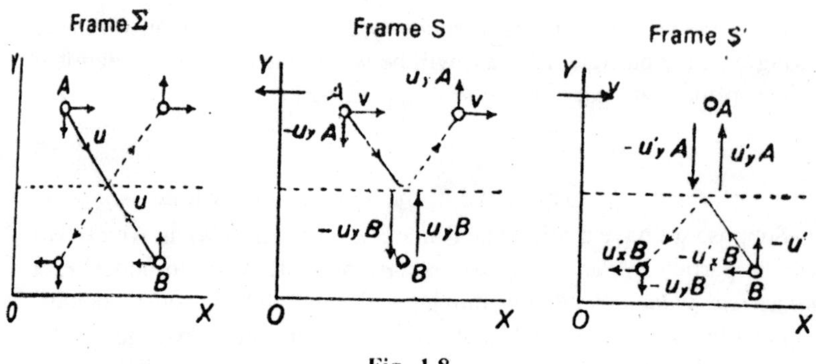

Fig. 1.8

In view of the initial symmetry of the glancing collision, the situations in frames S and S also naturally symmetrical, with particles A and B interchanging their roles. We, therefore, have

$$u'_y + A = u_y B. \qquad \text{...(i)}$$

In accordance with Lorentz transformation,

$$u_y A = \frac{u'_y A}{\gamma(1 + vu'_x A/c^2)},$$

where $u'_x A$ is the horizontal or x-component of A in frame S', which is *zero*, in this case.

So that, $\quad u_y A = u'_y + A/\gamma = u_y B/\gamma$

Or, $u_yA = u_yB/\gamma = u_yB\sqrt{1 - v^2/c^2}$,

$$\left[\because \gamma = 1/\sqrt{1-y^2/c^2}\right.$$

showing that the vertical components of the velocities of particles A and B, which are equal in magnitude in the stationary frame Σ, are no longer so in the moving frame S.

Considering the position in frame S, we find that the x-component of particle A remaining unchanged in magnitude and direction after collision, and particle B having *zero x-component, there is no change of momentum along the x-direction*, the only change thus occurring along the y-direction.

Clearly, *change in the vertical velocity-component of particle A, on collision* $= U_yA - (-u_yA) = 2u_yA$,

and change in the vertical velocity-component of particle B,

$$= -u_yB - u_yB = -2u_yB.$$

If, therefore, m_0 be the mass of each particle (supposed invariant, in accordance with Newtonian physics), we have

Change in momentum of the system $= 2m_0u_yA - 2m_0u_yB$,

which is obviously *not equal to zero*, since $u_yA \neq u_yB$, as we have seen above.

In order, therefore, that the law of conservation of momentum should hold, as it must, we must have the masses of particles A and B in the moving frame S different from their *Newtonian* or *rest masses* in the stationary frame S.

Let these be m_A and my respectively.

Then, for the law of conservation of momentum to hold, we must have $2m_Au_yA - 2m_Bu_yB = 0$.

Or, $m_Au_yA = m_BuyB$.

Or, since $u_yA = u_yB/\gamma$ and, therefore, $u_yB = \gamma u_yA$, we have

Or, $m_Au_yA = \gamma m_Bu_yA$, whence, $m_A = \gamma m_B$.

Or, $$M_A = \frac{m_B}{\sqrt{1 - v^2/c^2}}$$

Now, if in frame S' the vertical components u_yA and u_yB be small compared with the horizontal velocity-component v of particle A, we practically have particle B at rest and particle A moving relative to it with velocity v. Taking the rest mass of particle, therefore, to be m_0B, we have

$$M_A = \frac{m_0B}{\sqrt{1 - v^2/c^2}}$$

But, as we know, the two particles being *identical* in every respect, $m_0B = m_0A$, and, therefore,

$$M_A = \frac{m_0A}{\sqrt{1 - v^2/c^2}}$$

Or, representing the rest mass (or inertial mass) of a body, in general, by m_0 and its mass (or, relativistic mass), when moving with velocity v, by m, we have

$$m = \frac{m_0}{\sqrt{1 - v^2/c^2}} = \gamma m_0$$

which is the *mass-velocity relation of the Special theory of relativity*, indicating that *mass is no longer the absolute, invariant quantity c.f Classical physics but depends upon the velocity with which it is moving.*

As m the case of *length* and *time*, so also here, however, this relativistic effect in appreciable only at speeds somewhat comparable with that of light, *e.g.*, those of atomic particles. In case v is small, as it normally is, so that $v/c \rightarrow 0$, we have $m = m_0$, *i.e.*, the mass of the body is then the same as its rest mass, a result in accord with the Classical physics.

The first experimental confirmation of the relativity of mass came from *Bucherer*, in the year 1908, when he showed that the value of e/m (*ratio of charge to mass*) for fast moving electrons was smaller than for slower moving once, due obviously to a higher value of m in the former case, the value of e being always invariant.

Again, to deflect high speed electrons in a synchroton, the magnetic field required is 2000 times as strong as the one that would serve the purpose on the basis of Newton's laws, indicating clearly that the mass of the electrons has, in the synchroton, increased to 2000 times their normal mass, or, in other words, an electron now has the mass of a proton!

The relativistic relationship for mass, together with those for length and time stand fully confirmed and accepted as important basic formulations in atomic physics. The Newton's law of conservation of momentum, $p = m_0 v$, where m, is the rest mass of the body, supposed invariant in all frames of reference, is not invariant to Lorentz transformation and cannot, therefore, be acceptable as the correct law at relativistic velocities. What remains constant at these high velocities and is invariant to Lorentz transformation is $\gamma m_0 v$ or $m_0 v\ (1-v^2/c^2)^{1/2}$. So that, the relativistic law of conservation of momentum is $p = \gamma/m_0 v$.

At nonrelativistic velocities, $v/c \to 0$ and therefore, $\gamma = 1$, so that we have $p = m_0 v$, the Newtonian law of conservation of momentum, *i.e.*, at smaller velocities compared with c, the relativistic law of conservation of momentum reduces to the Newtonian law. The relativistic law is, therefore, accepted now as the correct law of conservation of momentum, invariant in all reference frames, the Newtonian law being a particular case of it. For most of our ordinary purposes, however, $v << c$ and hence the Newtonian law is good enough. It is only when we deal with high speed particles, like *electrons*, *protons*, *mesons*, etc., that we have to invoke the help of the *Relativistic law*.

If we were to plot the value of momentum p of a particle against velocity v, we should, according to the Newtonian or the Classical theory, obtain a straight line passing through the origin and the momentum should go on increasing with v even if v exceeds c, as shown by the dotted curve in Fig. 1.9.

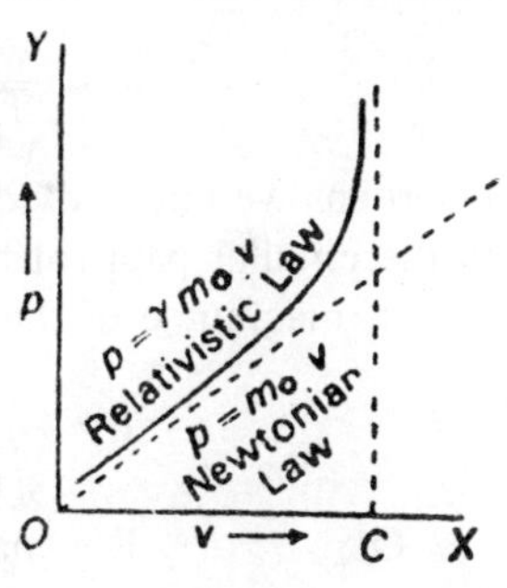

Fig. 1.9

On the other hand, if we plot p against v in accordance with the relativistic law of conservation of momentum, we obtain the full line curve shown, which shows that as $v \to c$, $p \to \infty$, because the mass then tends to infinity.

Force : In classical mechanics, as we know, force is defined as the *rate of change of momentum*, *i.e.*, $F = dp/dt$. This applies equally well to relativistic mechanics. Since mass here means relativistic mass, the expression for force, in terms of acceleration works out to be different from the classical expression $F = ma = mdv/dt$ in the case of linear *motion*, for,

$$F = \frac{dp}{dt} = \frac{d}{dt}(mv) = \frac{d}{dt}\left(\frac{m_o v}{\sqrt{1 - v^2/c^2}}\right)$$

Now, in this type of motion, *only the magnitude of the velocity changes and not its direction.* We may, therefore, put the above relation for force in the form

$$F = \frac{d}{dt}\left(\frac{m_o v}{1 - v^2/c^{1/2}}\right) = \frac{m_o (dv/dt)}{(1 - v^2/c)^{3/2}}$$

Or, since $m_o/(1 - v^2/c^2)^{1/2} = m$, we have

$$F = \frac{m}{(1 - v^2/c^2)}\frac{dv}{dt},$$

where dv/dt is, obviously, the rate of change of velocity or the acceleration of the body.

This, it will readily be seen, is different from the Classical relation $F = ma = mdv/dt$.

In the case of *circular motion*, however, the direction of the velocity of the body changes continuously, its magnitude remaining constant. So that, expression I above takes the form

$$F = \frac{m_o}{\sqrt{1 - v^2/c^2}} \cdot \frac{dv}{dt} = mdv/dt,$$

where, as we know, dv/dt is the *centripetal acceleration* (v^2/r). normal to the circular path (of radius r) of the body.

∴ magnitude of the force is given by

$$F = \frac{m_o}{\sqrt{1 - v^2/c^2}} \cdot \frac{v^2}{r} = \frac{mv^2}{r}.$$

Or, $F = ma$,

the same relation as in the case of Classical mechanics.

Thus, in the case of circular motion, the Relativistic formula for force is identical with the Classical one ($F = ma$), if only we use relativistic mass m in place of the classical mass.

MICHELSON—MORLEY EXPERIMENT

The apparatus used by Michelson and *Morley* is known as *Interferometer* since it depends upon the principle of interference of light and is shown diagrammatically in Fig. 1.10.

Here, S is a source of monochromatic light, a parallel beam from which falls upon a thin, parallel-sided glass plate P_1, thinly silvered on the back surface. The incident beam is thus split up into two parts at A at right angles to each other, and of *equal intensity, viz.*, a reflected one, travelling upwards and a refracted one, travelling along the original direction. The former suffers reflection at mirror M_1 at B along its own path and gets refracted through P_1 on to the telescope T. The latter similarly gets reflected back as it falls normally on mirror M_2, at C and is then reflected downwards from the back surface of P_1 along the same path as the former to enter the telescope. The two mirrors (M_1 and M_2) are heavily silvered on the front face (to avoid multiple reflections) and are arranged at right angles to each other at the same distance D from plate P_1. As will be easily seen, the beam reflected upwards to M_1 traverses the thickness of plate P_1 *thrice* whereas the one refracted on to mirror M_2 does so only *once*. To make their paths through glass and air equal, therefore, a *compensating plate* P_2, identical with P_1, is arranged parallel to it, as shown.

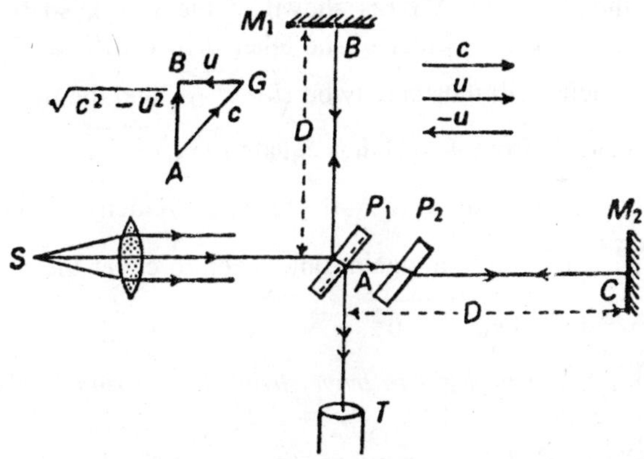

Fig. 1.10

If the beam of light be exactly parallel and if distances AB and AC from plate P_1 to the two mirrors respectively be the same (D), the two parts of the beam from M_1 and M_2 arrive at the telescope *in phase* and the field of view appears bright. If, however, their paths differ by an odd number of half wavelengths, they arrive at the telescope *out of phase, i.e.*, in opposite phases, and the field of view appears dark.

Since the beam of light incident on P_1 is almost always a slightly divergent one, some of the rays reaching the telescope from M_1 and M_2 are in phase and some out of phase, so that what we observe through the telescope is an *interference pattern of dark and bright fringes.*

Now, if u be the velocity of the apparatus (*i.e.*, of the earth), relative to the ether, from left to right, the drift velocity of the ether must be – *m, i.e., in the opposite direction, from right to left.* So that, if c be the actual velocity of light its relative velocity with respect to the apparatus, *along* AC = (c – u) and, therefore, time taken to cover the distance D from A to C, say, $t_1 = D/(c - u)$. And, on the return journey, from C to A, the relative velocity of light = (c + u) and, therefore, time taken to cover the same distance D from C to A, say, $t_2 = 2)/(c + u)$.

∴ total time taken by light to travel from A to C and back, say, t, is given by

$$t = t_1 + t_2 = \frac{D}{(c-u)} + \frac{D}{(c+u)} = \frac{2cD}{(c^2 - u^2)} = \frac{2D/C}{1-u^2/c^2}.$$

And the beam proceeding upwards from A to 5 must obviously be moving along the direction AG (as shown in the inset), so that the resultant of its velocity c and that of the ether drift (–u) is along AB. This resultant velocity will thus clearly be $\sqrt{c^2 - u^2}$ and the time taken to cover the distance D from A to B thus equal to $D/\sqrt{c^2 - u^2} = t_1$, say. On the return journey too from B to A, the resultant velocity of light will be the same $\sqrt{c^2 - u^2}$, so that, again, time taken to cover the distance D from B to A = $t_1' = D/\sqrt{c^2 - u^2}$.

∴ *total time taken by light to travel from A to B and back say,*

$$t' = t_1' + t_1' = 2t_1' = \frac{2D}{\sqrt{c^2 - u^2}} = \frac{2D/c}{\sqrt{1 - u^2/c^2}}.$$

As will be readily seen. the numerators in the expressions for t and t' are the *same, viz.*, 2D/c, and represent the time that light would take to cover the distance 2D, from A to C or from A to B and back, *if the apparatus were at rest.* In the denominator, the term u^2/c^2 is very small unless u happens to be comparable with c. Clearly, therefore, time t' taken by light from A to B and back is a little less than time t, taken by it from

A to C and back, even though the distance covered if the same D in either case.

This difference in time

$$\Delta t = t - t' = \frac{2D/c}{1-\frac{u^2}{c^2}} - \frac{2D/c}{\sqrt{1-\frac{u^2}{c^2}}} = \frac{2D}{c}\left(1-\frac{u^2}{c^2}\right)^{-1} - \frac{2D}{c}\left(1-\frac{u^2}{c^2}\right)^{-1/2}$$

Or, expanding by Binomial theorem and taking u < <c, we have

$$\Delta t = \frac{2D}{c}\left[\left(1+\frac{u^2}{c^2}\right)-\left(1+\frac{1}{2}\frac{u^2}{c^2}\right)\right] = \frac{2D}{c}\left(\frac{u^2}{2c^2}\right) = \frac{Du^2}{c^3}.$$

And, therefore, distance covered by light in time Δt

$$= \Delta t \times c = \frac{Du^2}{c_3} c = D\frac{u^2}{c^2},$$

indicating that the *optical path AC* is longer by Du^2/c^2 than the optical path AB or that this is the *path difference introduced* between the two parts of the incident beam (reflected from M_1 and M_2 respectively) *due to motion of the apparatus.*

The apparatus is now turned through 90° so that AB comes into the line of motion and AC, perpendicular to it; so that the optical path AB is now longer than the optical path AC by the same amount Du^2/c^2. The total path difference introduced between the two beams is thus ZDu^2/c^2. Naturally, therefore, the interference pattern in the field of view of the telescope will shift a little, say, through n fringes. Since for a path difference equal to one wavelength. λ, the pattern shifts through 1 fringe, its shift through n fringes indicates r path difference $n\lambda$. We, therefore, have

$$n\lambda = 2Du^2/c^2, \quad \text{whence, } n = 2Du^2/c^2\lambda.$$

Michelson and *Morley* had effectively increased distance D to nearly 11 metres (by repeated reflections) and, in order that the apparatus may be rotated without producing any strain, it was mounted on a stone slab, floated in mercury and kept rotating slowly at about 10 rotations per hour.

Thus, with *orbital velocity u of the earth, and hence that of the apparatus*, equal to 18.5 *miles* or 30 km/sec or 3×10^6 cm/sec, the velocity of light, $c = 3 \times 10^{10}$ cm/sec and with the wavelength of light used, $\lambda = 6 \times 10^{-6}$ cm, the expected value of n, in accordance with the expression above, comes to

$$\frac{2 \times 1100 \times 9 \times 10^{12}}{9 \times 10^{20} \times 6 \times 10^{-5}} = 0.37 \approx 0.4,$$

which could be accurately measured, in view of the high sensitivity of the apparatus, capable of measuring a shift as small as one-hundredth of a fringe.

The actual shift of the interference pattern observed, however, was much too small, almost negligible, *indicating little or no relative velocity between the earth and the ether.*

The experiment has since been repeated below the earth's surface as well as at high altitudes at different times of the year and, lately, with highly monochromatic light from a *laser*, but in all cases, the result has been a negative one. Other experiments to discover the ether wind have similarly failed.

It is always hard, however, to discard an idea to which one has grown accustomed. Strenuous efforts were therefore made to explain away the negative result of the experiment. Thus, Michelson himself hazardous the suggestion that the earth dragged along with it the ether in its immediate neighbourhood and there was thus no relative motion between the two. This was, however, easily demolished by *Lodge* who, in the year 1892, measured the velocity of light near rapidly rotating bodies and came to the conclusion that not more than half a per cent of the velocity of light could thus be communicated to the ether.

The same vear (1892), *Fitzgerald* and *Lorentz* put forth the suggestion that probably there was interaction between the ether and a material body moving relatively to it, and that, as a result, the body gets shortened in all its dimensions *parallel to the relative velocity*, such that if L_0 be the length of the body when at rest, and if it be moving with a speed u parallel to its length, the new length L acquired by it is given by the relation, $L = L_0 \sqrt{1 - u^2/c^2}$. It will be easily seen that if we make use of this suggestion in the experiment discussed above, we shall find that whereas distance AB will remain unchanged, distance AC will get shortened to $D\sqrt{1 - u^2/c^2}$. So that, substituting $D\sqrt{1 - u^2/c^2}$ for D in the expression for t above, we have

$$t = \frac{(2D/c)\left(\sqrt{1 - u^2/c^2}\right)}{1 - u^2/c^2} = \frac{2D/c}{\sqrt{1 - u^2/c^2}}, \text{ which is the same as } t'.$$

This contraction hypothesis, therefore, easily explains why the *Michelson-Morley Experiment* gave a negative result. It is, however, open to the objection that it has been put forward for the specific purpose of explaining away the difficulty and that it does not follow from the theory.

It was 18 year's later, in 1905, that the true and simple explanation for the negative result of Michelson and Morley's experiment was furnished by *Einstein*, namely, that *the velocity of light in space is* a *universal constant*. In fact, he made this constancy of the velocity of light one of the basic postulates of his special theory of relativity.

SOLVED EXAMPLES

Example 1:

How much energy will be obtained if 3.6 g of mass be completely converted into energy?

Solution:

By mass-energy relation, we have

$$\Delta E = (\Delta m)c^2.$$

The mass annihilated, $\Delta m = 3.6$ g $= 3.6 \times 10^{-3}$ kg and speed of light $c = 3 \times 10^8$ m/s.

$$\therefore \quad \Delta E = (3.6 \times 10^{-3})(3.0 \times 10^8)^2$$

$$= 32.4 \times 10^{13} \text{ J}.$$

Now 1 MeV $= 1.60 \times 10^{-13}$ J.

$$\therefore \quad \Delta E = \frac{32.4 \times 10^{13}}{1.6 \times 10^{-13}} = 2.0 \times 10^{27} \text{ MeV}.$$

Example 2:

If 1 kg of mass be completely converted into energy, how much kilowatt-hour electricity would be obtained from it?

Solution:

The energy obtained is given by

$$\Delta E = (\Delta m)\, c^2$$

$$= (1 \text{ kg})\, (3.0 \times 10^8 \text{ m/s})^2$$

$= 9.0 \times 10^{16}$ J.

Now 1 kilowatt-hour = (1000 × 3600) W-s = 3.6×10^6 J.

∴ electricity obtained is

$$\frac{9.0 \times 10^{16}}{3.6 \times 10^6} = 2.5 \times 10^{10} \text{ kW-h.}$$

Example 3:

What is the mass equivalent of the energy from an antenna radiating 10000 watts for 24 hours?

Solution:

The total energy radiated is

$$\Delta E = 10000 \times 24 \text{ watt-hours}$$

$$= 10000 \times 24 \times 3600 \text{ watt-sec}$$

$$= 8.64 \times 10^8 \text{ joule.}$$

By Einstein's relation, the mass equivalent is

$$\Delta m = \frac{\Delta E}{c^2}$$

$$= \frac{8.64 \times 10^8 \text{ joule}}{(3.0 \times 10^8 \text{ m/s})^2} = 9.6 \times 10^{-9} \text{ kg.}$$

Example 4:

In Michelson-Morley experiment the lengths of the paths of the two beams is 11 meter each. The wavelength of the light used is 6000 Å. If die expected fringe-shift is 0.4 fringe, calculate the velocity of earth relative to ether.

Solution:

The expected shift in the Michelson-Morley experiment is given by

$$\Delta N = \frac{2l v^2}{c^2 \lambda}.$$

Therefore, the velocity of earth relative to ether is

$$v = c\sqrt{\frac{\lambda \Delta N}{2l}}.$$

Here l = 11 m, l = 6000 Å = 6×10^{-7} m and ΔN = 0.4.

$$\therefore\ v = (3.0 \times 10^8 \text{ m/s}) \sqrt{\frac{(6 \times 10^{-7} \text{ m}) \times 0.4}{2 \times 11 \text{ m}}}$$

$$= (3.0 \times 10^8 \text{ m/s})(1.044 \times 10^{-4})$$

$$= 3.1 \times 10^4 \text{ m/s.}$$

Example 5:

An event occurs at $x = 100$ km $y = 10$ km and $z = 1.0$ km at $t = 2.0 \times 10^{-4}$ second in a reference frame S. Another frame S' is moving with speed 0.95 c relative to S along the common $X - X'$ axis, the origins coinciding at $t = t' = 0$. Compute the coordinates x', y', z', t' of the event in S'.

Solution:

By Lorentz transformation, we have

$$x' = \frac{x - vt}{\sqrt{1 - \left(\frac{v}{c}\right)^2}} = \frac{100 \text{ km} - (0.95 \times 3 \times 10^5 \text{ km/s})(2.0 \times 10^{-4} \text{ s})}{\sqrt{1 - (0.95)^2}}$$

$$= \frac{100 \text{ km} - 57 \text{ km}}{\sqrt{1 - 0.9025}} = \frac{43 \text{ km}}{\sqrt{0.0975}} = \frac{43}{0.312} = 137.8 \text{ km}.$$

$y' = y = 10$ km.

$z' = z = 1.0$ km.

$$t' = \frac{t - \left(\frac{xv}{c^2}\right)}{\sqrt{1 - \left(\frac{v}{c}\right)^2}} = \frac{(2.0 \times 10^{-4} \text{ s}) - \frac{100 \text{ km} \times 0.95}{3 \times 10^5 \text{ km/s}}}{\sqrt{1 - (0.95)^2}}$$

$$= \frac{2.0 \times 10^{-4} \text{ s} - 3.166 \times 10^{-4} \text{ s}}{0.312} = -3.74 \times 10^{-4} \text{ s}.$$

Example 6:

The mass of deuteron is less than the sum of the masses of one proton and one neutron by 0.0024 amu where 1 amu = 1.67×10^{-27} kg. Deduce the energy given off (as γ-radiation) when a neutron and a proton combine to form a deuteron. How does the case of fission of U^{235} nucleus differ from this case?

Solution:

The mass disappeared during the formation of deuteron is

$$\Delta m = 0.0024 \text{ amu} = 0.0024 \times (1.67 \times 10^{-27}) \text{ kg}$$
$$= 0.0040 \times 10^{-27} \text{ kg.}$$

The corresponding energy given off is

$$\Delta E = \Delta m\, c^2 = (0.0040 \times 10^{-27} \text{ kg}) \times (3 \times 10^8 \text{ m/s})^2$$
$$= 3.6 \times 10^{-13} \text{ joule.}$$

In this case the released energy goes as γ-rays. In nuclear fission a heavy nucleus (such as U^{235}) is splitted into lighter nuclei. In this process, the binding energy per nucleon increases and results in a general release of huge about of energy. Most of the energy released goes into the kinetic energy of the fission fragments, while about 20% goes as neutrons, γ-rays, neutrinos, etc.

Example 7:

In a reference frame S an event 1 occurs at the origin at $t = 0$ and another event 2 occurs at $x = 3000$ meter and $t = 4 \times 10^{-6}$ second. Find the time-interval between events as registered by clocks in a frame S' moving with a speed $v = 0.6\,c$ relative to 5 along the common $X - X'$ axis, the origins coinciding at $t = t' = 0$.

Solution:

Let t_1' and t_2' be the times of the events 1 and 2 as registered on S' clocks. Then, by Lorentz transformation, we have

$$t_1' = \frac{t_1 - \left(\frac{x_1 v}{c^2}\right)}{\sqrt{1 - \left(\frac{v^2}{c^2}\right)}} = 0$$

and $$t_2' = \frac{t_2 - \left(\frac{x_2 v}{c^2}\right)}{\sqrt{1 - \left(\frac{v}{c}\right)^2}} = \frac{4 \times 10^{-6}\text{s} - \frac{3000\text{m} \times 0.6}{3 \times 10^8 \text{ m/s}}}{\sqrt{1 - (0.6)^2}}$$

$$= \frac{4 \times 10^{-6} - 6 \times 10^{-6}}{0.8} = -2.5 \times 10^{-6}\text{s}.$$

$\therefore$ time-interval $t_2' - t_1' = -2.5 \times 10^{-6}$ s.

The S' clock registers the event 2 earlier than S clock.

Example 8:

A rocket ship is 100 meters long on the ground. When it is in flight its length is 99 meters to an observer on the ground. What is its speed? ($c = 3 \times 10^8$ m/s).

Solution:

Let L_0 be the length of the rocket ship on the ground. When it is in flight with velocity v, it appears contracted to a length L to an observer on the ground, where

$$L = L_0\sqrt{1-\left(\frac{v^2}{c^2}\right)},$$

where c is the speed of light. From this, we have

$$1-\frac{v^2}{c^2} = \left(\frac{L}{L_0}\right)^2$$

or
$$v^2 = \left[1-\left(\frac{L}{L_0}\right)^2\right]c^2$$

$\therefore$
$$v = \sqrt{1-\left(\frac{L}{L_0}\right)^2}\,c.$$

Here $\frac{L}{L_0} = \frac{99}{100} = 0.99$.

Now, $\sqrt{1-0.99^2} = \sqrt{(1-0.99)\times(1+0.99)} = \sqrt{0.01\times 1.99}$

$$= \sqrt{0.0199} = 0.141.$$

$\therefore$ $v = 0.141$ c.

Example 9:

The proper mean life-time of π meson is 2.5×10^{-8} second. Calculate the mean life-time of a π meson travelling with a velocity of 2.4×10^{10} cm/s, distance travelled by it before disintegrating, distance it would travel if there were no relativity effect.

Solution:

The life-time of a moving particle appears to be lengthened to a stationary observer. If t_0 be the actual half-life (in the frame of reference of the particle itself) and t that measured by a stationary observer, then

$$t = \frac{t_0}{\sqrt{1-\left(\frac{v^2}{c^2}\right)}}.$$

Here $t_0 = 2.5 \times 10^{-8}$ s, $v = 2.4 \times 10^{10}$ cm/s and we know that $c = 3.0 \times 10^{10}$ cm/s.

$$\therefore \quad t = \frac{2.5 \times 10^{-8}\,\text{s}}{\sqrt{1-\left(\frac{2.4 \times 10^{10}}{3.0 \times 10^{10}}\right)}} = \frac{2.5 \times 10^{-8}\,\text{s}}{\sqrt{1-(0.8)^2}}$$

$$= \frac{2.5 \times 10^{-8}\,\text{s}}{0.6} = 4{,}17 \times 10^{-8}\,\text{s}.$$

Distance travelled by the particle before disintegrating

= velocity × life-time

= $(2.4 \times 10^{10}$ cm/s$) \times (4.17 \times 10^{-8}$ s)

= 1000cm.

Distance travelled if there were no relativity effect

= velocity × proper life-time

= $(2.4 \times 10^{10}$ cm/s$) \times (2.5 \times 10^{-8}$ s$)$ = 600 cm .

Example 10:

Calculate the speed of an electron which has kinetic energy 1.02 MeV. The rest mass of the electron is 0.51 MeV. (c = 3 × 10^{10} cm/s).

Solution:

The relativistic expression for the kinetic energy of a particle is m_0c^2 $\left\{\frac{1}{\sqrt{1-(v/c)^2}} - 1\right\}$. When an electron acquires it under a potential of V volts, we have

$$m_0c^2\left\{\frac{1}{\sqrt{1-\left(\frac{v}{c}\right)^2}}-1\right\}=eV$$

so that

$$\frac{1}{\sqrt{1-\left(\frac{v}{c}\right)^2}}=1+\frac{eV}{m_0c^2}.$$

The rest mass of the electron is given in energy units (electron volts); this means that we are given the rest mass *energy* (m_0c^2) of the electron. Thus, here

$$eV = 1.02 \text{ MeV}$$

and $m_0c^2 = 0.51$ MeV.

$$\therefore \quad \frac{1}{\sqrt{1-\left(\frac{v}{c}\right)^2}}=1+\frac{1.02 \text{ MeV}}{0.51 \text{ MeV}} = 1+2=3$$

or $$1-\frac{v^2}{c^2}=\left(\frac{1}{3}\right)^2=\frac{1}{9}$$

or $$\frac{v^2}{c^2}=1-\frac{1}{9}=\frac{8}{9}$$

or $$\frac{v}{c}=0.943.$$

$$\therefore \quad v = 0.943\ c = 0.943 \times (3 \times 10^{10})$$

$$= 2.83 \times 10^{10} \text{ cm/s}.$$

Example 11(a):

An electron (rest mass 9.1×10^{-31} kg) is moving with speed 0.99 c. What is its total energy? Find the ratio of Newtonian kinetic energy to the relativistic energy. ($c = 3.0 \times 10^8$ m/s).

Solution:

The mass of the electron (rest mass $m_0 = 9.1 \times 10^{-31}$ kg) moving with velocity v (= 0.99 c) is

$$m = \frac{m_0}{\sqrt{1-\left(\frac{v}{c}\right)^2}} = \frac{9.1 \times 10^{-31}}{\sqrt{1-(0.99)^2}}$$

$$= \frac{9.1 \times 10^{-31}}{\sqrt{1-0.98}} = \frac{9.1 \times 10^{-31}}{0.141} = 64.5 \times 10^{-31} \text{ kg.}$$

The total energy of the electron is therefore

$$E = mc^2 = (64.5 \times 10^{-31} \text{ kg})(3.0 \times 10^8 \text{ m/s})^2 = 5.8 \times 10^{-13} \text{ J.}$$

The Newtonian kinetic energy os $1/2\ m_0v^2$, while the relativistic kinetic energy is total energy *minus* rest energy. Thus, their ratio is

$$\frac{(1/2)m_0c^2}{mc^2 - m_0c^2} = \frac{1}{2}\frac{m_0}{m-m_0}\left(\frac{v}{c}\right)^2$$

$$= \frac{1}{2}\frac{9.1 \times 10^{-31}}{55.4 \times 10^{-31}}(0.99)^2 = 0.08.$$

Example 11(b):

A beam of particles of half-life 2.8, × 10 ⁶ second travels in the laboratory with speed 0.96 times the speed of light. How much distance does the beam travel before the flux falls to one-half time the initial flux? (c = 3.0 × 10⁸ m/s).

Solution:

The proper half-life of the beam of particles is $t_0 = 2.8 \times 10^{-6}$ s. The apparent half-life of the beam moving with velocity v as measured by a stationary observer is

$$t = \frac{t_0}{\sqrt{1-\left(\frac{v^2}{c^2}\right)}}$$

$$= \frac{2.8 \times 10^{-6}}{\sqrt{1-(0.96)^2}} = \frac{2.8 \times 10^{-6}}{\sqrt{(1-0.96)(1+0.96)}}$$

$$= \frac{2.8 \times 10^{-6}}{\sqrt{0.04 \times 1.96}} = \frac{2.8 \times 10^{-6}}{0.28} = 1 \times 10^{-5} \text{s.}$$

The half-life is defined to be the time during which half of the initial particles decay. Thus, the distance travelled before the flux falls to half

= speed × half-life (apparent)

= 0.96 c × t

= $0.96 \times (3 \times 10^8 \text{ m/s}) \times (1 \times 10^{-5} \text{ s})$

= 2.88×10^3 m = 2.88 km.

Example 12:

What is the velocity of π-mesons whose proper mean life is 2.5×10^{-8} second and observed mean life is 2.5×10^{-7} second?

Solution:

As a result of relativistic time dilation, the observed mean life is given by

$$t = \frac{t_0}{\sqrt{1-\left(\frac{v}{c}\right)^2}},$$

where t_0 is the proper life and v the speed of the particle. Thus

$$\sqrt{1-\left(\frac{v}{c}\right)^2} = \frac{t_0}{t} = \frac{2.5 \times 10^{-8}}{2.5 \times 10^{-7}} = 0.1$$

or $$1 - \frac{v^2}{c^2} = 0.01$$

or $$\frac{v^2}{c^2} = 1 - 0.01 = 0.99$$

or $$v = \sqrt{0.99}\ c = 0.995\ c.$$

Example 13(a):

A meson has a speed 0.8 c relative to the ground. Find how far the meson travels relative to the ground, if its speed remains constant and the time of flight, relative to the system in which it is at rest, is 2.0×10^{-8} second. ($c = 3.0 \times 10^8$ m/s).

Solution:

If t_0 be the time of flight of the meson in its own frame of reference, then the time of flight relative to the ground is

$$t = \frac{t_0}{\sqrt{1-\left(\frac{v^2}{c^2}\right)}}$$

$$= \frac{2.0 \times 10^{-8}\,s}{\sqrt{1-(0.8)^2}} = 3.33 \times 10^{-8}\,s$$

∴ distance travelled relative to the ground is

speed × time = 0.8 c × (3.33 × 10^{-8} s)

= (0.8 × 3.0 × 10^8 m/s) × (3.33 × 10^{-8} s)

= 8.0 meter.

Example 13(b):

A clock keeps correct time. With what speed should it be moved relative to an observer so that it may appear to lose 4 minutes in 24 hours?

Solution:

If to be the (proper) time-interval on clock at rest relative to an observer and t the time-interval on clock in motion relative to the observer, then

$$t = \frac{t_0}{\sqrt{1-\left(\frac{v}{c}\right)^2}}$$

where v is the speed of relative motion.

Here t_0 = 24 × 60 = 1440 min is the proper time-interval measured by an observer moving with the clock and t = 1444 min is the same time-interval measured (with his own clock) by a stationary observer. Thus, to the stationary observer, the moving clock appears to lose 4 min in 24 hours. From the above formula,

$$1444 = \frac{1440}{\sqrt{1-\left(\frac{v}{c}\right)^2}}$$

For $\frac{v}{c} << 1$, $\frac{1}{\sqrt{1-\left(\frac{v^2}{c^2}\right)}} \approx 1+\frac{v^2}{2c^2}$, then

$$1444 = 1440\left(1 + \frac{v^2}{2c^2}\right)$$

or $$\frac{v^2}{2c^2} = \frac{1444}{1440} - 1 = \frac{4}{1440}$$

or $$\frac{v^2}{c^2} = \frac{8}{1440} = 0.0055$$

or $$v = \sqrt{0.0055}\,c = 0.745\ c$$

$$= (0.0745)\ (3 \times 10^8\ \text{m/s})$$

$$= 2.23 \times 10^7\ \text{m/s}.$$

Example 14:

Two particles are moving in opposite directions each with a speed of 0.9 c in laboratory frame of reference. Find the velocity of one particle relative to other.

Solution:

The particles are moving with velocities + 0.9 c and – 0.9 c in the laboratory frame, say S' frame. Let S be a reference frame in which the particle with velocity – 0.9 c is at rest. Then the velocity of S' (laboratory) relative to S is v = 0.9 c.

Therefore, the particle which in S' has velocity u' = + 0.9 c has a velocity in S given by

$$u = \frac{u' + v}{1 + \frac{u'v}{c^2}}$$

$$= \frac{+0.9c + 0.9c}{1 + \frac{(0.9c)\,(0.9c)}{c^2}}$$

$$= \frac{1.8c}{1 + (0.9)^2} = \frac{1.8c}{1.81} = 0.994\ c,$$

where is less than c. If the Galilean velocity transformations were used, we would find that u = u + v = 0.9 c + 0.9 c = 1.8 c , which is greater than c and hence impossible.

Example 15:

Rockets A and B are observed from the earth to be travelling with velocities 0.8 c and 0.7 c in the same line in the same direction. What is the velocity of B as seen by an observer on A?

Solution.

Rockets A and B are moving with velocities 0.8 c and 0.7 c in the earth's frame, say S' frame. Let S be a frame in which A is at rest. Then, the velocity of S' (earth) relative to S is v = - 0-8 c . Therefore, rocket B which in S' has velocity u' = 0.7 c has a velocity in S (observer on A) given by

$$u = \frac{u'+v}{1+\frac{u'v}{c^2}}$$

$$= \frac{0.7c+(-0.8c)}{1+\frac{(0.7c)(-0.8c)}{c^2}}$$

$$= -\frac{0.1c}{1-0.56} = -0.23\ c.$$

The negative sign indicates that the rocket B is approaching A with time.

Example 16:

A spaceship moving away from the earth with velocity 0.5 c fires a rocket whose velocity relative to spaceship is 0.8 c (a) away from the earth, (b) towards the earth. What will be the velocity of the rocket as observed from the earth in the two cases.

Solution:

We may regard earth as the S-frame, the spaceship as the S'-frame, and the rocket as the object whose velocity in the S-frame we seek. Then

$$u' = +0.8\ c, \qquad v = 0.5\ c.$$

$$\therefore \quad u = \frac{u'+v}{1+\frac{u'v}{c^2}} = \frac{0.8c+0.5c}{1+(0.8\times 0.5)} = 0.93c$$

In the second case, u' = – 0.8 c.

$$\therefore \qquad u = \frac{u'+v}{1+\frac{u'v}{c^2}} = \frac{-0.8c+0.5c}{1+\{(0.8)\times(0.5)\}}$$

$$= -\frac{0.3c}{1-0.4} = -0.5c.$$

Example 17:

Find the resultant of a velocity of 0.6 c and a velocity of 0.8 c inclined at 60° to each other.

Solution:

Let us choose 0.6 c as velocity v along x-axis. Then

$$\gamma = \frac{1}{\sqrt{1-\left(\frac{v^2}{c^2}\right)}} = \frac{1}{\sqrt{1-(0.6)^2}} = \frac{1}{0.8}$$

The velocity 0.8 c can be resolved into a component along x-axis, given by

$$u_y' = (0.8\ c)\cos 60^\circ = 0.4\ c$$

and a component along y-axis, given by

$$u_y' = (0.8\ c)\sin 60^\circ = 0.693\ c.$$

By Lorentz transformation, the components of the resultant velocity are given by

$$u_x = \frac{u_x' + v}{1+\left(\frac{u_x' v}{c^2}\right)} = \frac{0.4c+0.6c}{1+(0.4\times 0.6)} = 0.806\ c$$

and
$$u_y = \frac{u_y'}{\gamma\left\{1+\left(\frac{u_x' v}{c^2}\right)\right\}} = \frac{0.693c}{\left(\frac{1}{0.8}\right)\{1+(0.4\times 0.6)\}} = 0.447c$$

The magnitude of the resultant velocity is

$$u = \sqrt{u_x^2 + u_y^2} = \sqrt{(0.806c)^2 + (0.447c)^2} = 0.922\ c.$$

It makes with 0.6 c or x-axis an angle given by

$$\theta = \tan^{-1}\left(\frac{u_y}{u_x}\right) = \tan^{-1}\left(\frac{0.447}{0.806}\right) = \tan^{-1}(0.55) = 29°.$$

Example 18:

On the surface of the earth the mass of a man is 100 kg. When he is in a rocket moving with a speed of 4.2×10^7 meters/second relative to the earth, what will be his mass as observed by (i) an observer on the earth, (ii) an observer in his rocket? ($c = 3.0 \times 10^8$ meters/second).

Solution:

(i) The rest mass of the man is $m_0 = 100$ kg. The mass as observed by a stationary observer when the man is moving in a rocket is

$$m = \frac{m_0}{\sqrt{1-\left(\frac{v}{c}\right)^2}}$$

$$= \frac{100 \text{ kg}}{\sqrt{1-\left(\frac{4.2\times10^7}{3.0\times10^8}\right)^2}} = \frac{100 \text{ kg}}{\sqrt{1-(0.14)^2}}$$

$$= \frac{100 \text{ kg}}{\sqrt{0.9804}} = 101 \text{ kg}.$$

(ii) An observer moving in the rocket will find the mass of the man the same as the rest mass, that is, 100 kg, because there is no relative motion between the observer and the man.

Example 19:

State the fundamental postulates of the special theory of relativity and deduce from them the Lorentz transformation equations.

Solution:

Einstein's Special Theory of Relativity : In 1905, Einstein published his special theory of relativity which is based upon two postulates.

Postulate I: *The laws of Physics have the same form in all inertial frames of reference moving with a 'constant' velocity relative to one another.* This postulate expresses the absence of a universal frame of reference. If the laws of Physics were different for observers in different frames in relative motion, it could be determined from these differences which objects are stationary in space and which are moving.

But because there is no universal frame of reference, this distinction between objects cannot be made. (If we are in a free balloon from which we cannot see anything except another free balloon which is changing its position relative to us, we cannot know which balloon is really moving).

Postulate II: *The speed of light in free space is the same in all inertial frames of reference.* This postulate follows directly from the result of the Michelson-Morley experiment.

Galilean-Newtonian Transformation Equations : Let S and S' (Fig. 1.11) be two frames of reference, S' moving along the + x direction with a constant velocity v relative to S. Let (x, y , z , t) be the space and time co-ordinates of an event for an observer in the frame S, and (x' , y', z', t') the coordinates of the same event for an observer in the frame S'. Let the time be counted from the instant when the origins O and O' momentarily coincide. We have to see how are the measurements x, y , z , t related to x', y', z', t'.

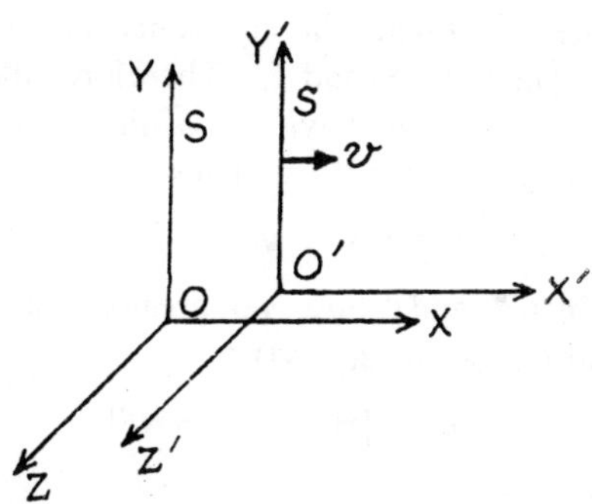

Fig. 1.11

According to the classical mechanics, the measurement in the x direction made in S' will be smaller than that made in S by the amount vt, which is the distance moved by S' in the x direction. That is

$$x' = x - vt. \qquad ...(i)$$

There is no relative motion in the y and z directions, and so

$$y' = y \qquad ...(ii)$$

and $z' = 2.$ ...(iii)

In classical mechanics, the time-intervals measured in different frames of reference in relative motion are the same, and so

$$t' = t. \qquad ...(iv)$$

The set of equations (i) to (iv) are the Galilean-Newtonian transformation equations. They, however, violate both of the postulates of special relativity. The equations of electromagnetism assume very different forms when these equations are used to convert quantities measured in one frame with the equivalent quantities in the other. This violates the I postulate. Again, if we measure the speed of light along the x direction in the frame S to be c, then in the frame S' it will be $c' = c - v$. This violates the II postulate.

Lorentz Transformation Equations : Let us now develop a set of transformation equations directly from the postulates of special relativity. We can reasonably assume that a measurement in the x direction made in S is proportional to that made in S'. That is, we may write

$$x' = \gamma(x - vt). \qquad ...(v)$$

According to the I postulate, the equations (laws) of Physics must have the same form in both S and S'. Therefore, the corresponding equation of x in terms of x' and t' will be of the same form as eq. (v), except that v will be replaced by – v. Thus

$$x = \gamma(x' + vt'). \qquad ...(vi)$$

The time coordinates t and t', however, are not equal. Let us substitute the value of x' from eq. (v) in eq. (vi) :

$$x = \gamma[\gamma(x - vt) + vt']$$

or
$$\frac{x}{\gamma} = \gamma x - \gamma vt + vt'.$$

$\therefore$
$$t' = \frac{x}{\gamma v} - \frac{\gamma x}{v} + \gamma t$$

$$= \gamma t - \frac{\gamma t}{v}\left(1 - \frac{1}{\gamma^2}\right). \qquad ...(vii)$$

Similarly, we can show that

$$t = \gamma t' + \frac{\gamma x'}{v}\left(1 - \frac{1}{\gamma^2}\right). \qquad ...(viii)$$

Now, γ can be evaluated from the II postulate. Suppose, a light signal is given at the origin O at time t = 0, t' = 0, *i.e.*, when O and O' coincide.

The signal travels with a speed c which is same for both the frames (II postulate). Its position as seen from S and S' after some time is given by

$$x = ct \quad \text{and} \quad x' = ct'.$$

Substituting for x and x' in eq. (v) and (vi), we obtain

$$ct' = \gamma t\,(c - v)$$

and $ct = \gamma t'\,(c + v)$.

Solving these two equations, we can easily see that

$$\gamma = \frac{1}{\sqrt{1-\left(\frac{v^2}{c^2}\right)}}. \qquad \text{...(ix)}$$

Inserting this value of γ in eq. (v) and (vi), we get

$$x' = \frac{x - vt}{\sqrt{1-\left(\frac{v^2}{c^2}\right)}} \qquad \text{...(x)a}$$

and

$$x = \frac{x' + vt'}{\sqrt{1-\left(\frac{v^2}{c^2}\right)}}. \qquad \text{...(x)b}$$

Again, inserting the value of γ in eq. (v) and (vi), we get

$$t' = \gamma t - \frac{\gamma x}{v}\left(\frac{v^2}{c^2}\right) = \gamma\left(t - \frac{xv}{c^2}\right)$$

or

$$t' = \frac{t - \left(\frac{xv}{c^2}\right)}{\sqrt{1-\left(\frac{v^2}{c^2}\right)}} \qquad \text{...(xi)a}$$

and similarly

$$t = \frac{t' + \left(\frac{x'v}{c^2}\right)}{\sqrt{1-\left(\frac{v^2}{c^2}\right)}}. \qquad \text{...(xi)b}$$

In the general case when the light signal is not restricted to travel along the x-axis, we get two more relations

$$y' = y \text{ and } z' = z, \qquad \text{...(xiii)}$$

since these distances are measured perpendicular to v.

Eq. (x), (xi) and (xii) are known as Lorentz transformation equations. They are used for the transformation of the space and time co-ordinates of an event from one frame of reference into the other in uniform relative motion.

There are two important aspects of the Lorentz equations:

(i) The measurements of position and time depend upon the frame of reference of the observer. Two events which are simultaneous when viewed from one frame of reference are not simultaneous when viewed from a second frame which is moving relative to the first.

(ii) The Lorentz equations reduce to the classical Galilean equations when $v << c$. Therefore, the consequences of special relativity are not apparent unless enormous velocities are involved.

Example 20:

(a) Show that $x^2 + y2 + z^2 - c^2t^2$ is invariant under Lorentz transformation.

(b) A light pulse is emitted at the origin of a frame of reference S' at time $t' = 0$. Its distance x' from the origin after a time t' is given by $x'^2 = c^2t'^2$. Use the Lorentz transformation to transform this equation to an equation in x and t and show that this is $x^2 = c^2t^2$. Discuss the implication of this result.

Solution:

(a) *Invariancy of $x^2 + y^2 + z^2 - c^2t^2$:* Let (x, y, z, t) and (x', y', z' t') be the space-time coordinates of an event observed by two observers in reference frame S and S' respectively, the frame S' moving with a constant velocity v relative to S . Using Lorentz transformation equations, we have to simply prove that

$$x^2 + y^2 + z^2 - c^2t^2 = x'^2 + y'^2 + z'^2 - c^2t'^2.$$

The Lorentz equations are

$$x' = \frac{x - vt}{\sqrt{1 - \left(\frac{v^2}{c^2}\right)}}, \; y' = y, \; z' = z, \; t' = \frac{t - \left(\frac{xv}{c^2}\right)}{\sqrt{1 - \left(\frac{v^2}{c^2}\right)}}$$

Then, we have

$x'^2 + y'^2 + z'^2 - c^2t'^2$

$$= \frac{(x-vt)^2}{1-\left(\frac{v^2}{c^2}\right)} + x^2 + z^2 - \frac{c^2\left\{t-\left(\frac{xv}{c^2}\right)\right\}^2}{1-\left(\frac{v^2}{c^2}\right)}$$

$$= \frac{1}{1-\left(\frac{v^2}{c^2}\right)}\left[x^2 + v^2t^2 - 2xvt - c^2\left(t^2 + \frac{x^2v^2}{c^4} - \frac{2t\,xv}{c^2}\right)\right] + y^2 + z^2$$

$$= \frac{1}{1-\left(\frac{v^2}{c^2}\right)}\left[x^2 + v^2t^2 - 2xvt - c^2t^2 - \frac{x^2v^2}{c^2} + 2txv\right] + y^2 + z^2$$

$$= \frac{1}{1-\left(\frac{v^2}{c^2}\right)}\left[x^2\left(1-\frac{v^2}{c^2}\right) - c^2t^2\left(1-\frac{v^2}{c^2}\right)\right] + y^2 + z^2$$

$$= [x^2 - c^2t^2] + y^2 + z^2 = x^2 + y^2 + z^2 - c^2t^2.$$

Hence $x^2 + y^2 + z^2 - c^2t^2$ is Lorentz-invariant.

(b) Let us consider the given equation

$$x'^2 = c^2t'^2.$$

By Lorentz transformation, we have

$$x' = \frac{x-vt}{\sqrt{1-\left(\frac{v^2}{c^2}\right)}} \quad \text{and} \quad t' = \frac{t-\left(\frac{xv}{c^2}\right)}{\sqrt{1-\left(\frac{v^2}{c^2}\right)}}.$$

Then, from the given equation, we have

$$\frac{(x-vt)^2}{1-\frac{v^2}{c^2}} = \frac{c^2\left\{t\left(\frac{xv}{c^2}\right)\right\}^2}{1-\frac{v^2}{c^2}}$$

or $$(x-vt)^2 = c^2\left(t^2 + \frac{x^2v^2}{c^4} - \frac{2t\,xv}{c^2}\right)$$

or $$x^2 + v^2t - 2xvt = c^2t^2 + \frac{x^2v^2}{c^2} - 2t\,xv$$

or $$x^2 - c^2t^2 + v^2t^2 - \frac{x^2v^2}{c^2} = 0$$

or $$\left(x^2 - c^2t^2\right)\left(1 - \frac{v^2}{c^2}\right) = 0.$$

Since $v \neq c$, $\left(1 - \frac{v^2}{c^2}\right) \neq 0.$

$\therefore$ $$x^2 - c^2t^2 = 0$$

or $$x^2 = c^2t^2.$$

This result shows that the velocity of light is an absolute constant independent of the frame of reference.

Example 21:

Explain (a) length contraction and (b) time dilation in special theory of relativity. What are proper lengtn and proper time-interval? How is time dilation experimentally verified?

Solution:

(a) *Lorentz-Fitzgerald Length Contraction :* Relative motion affects measurement of length. *A body moving with a velocity v relative to an observer appears to the observer to be contracted in length in the direction of motion by a factor* $\sqrt{1-\left(v^2/c^2\right)}$, *whereas its dimensions perpendicular to the direction of motion are unaffected.*

To prove this length contraction, let us consider a rod placed along the x'-axis of a moving frame of reference S'. Let x_1' and x_2' be the co-ordinates of the ends of the rod, as seen by an observer in S'. Then, the length of the rod as observed from S' is

$$L_0 = x_2' - x_1'.$$

L_0 is the rod's length as measured in a frame of reference in which it is at rest. Now, suppose another observer measures the length of the rod from a stationary frame of reference S, relative to which the rod (or the frame S') is moving with velocity v. If he records the co-ordinates of the ends of the rod as x_1 and x_2 at the time t, the rod appears to him as having a length

$$L = x_2 - x_1.$$

From the Lorentz transformation for a given value of t we have

$$x_2' = \frac{x_2 - vt}{\sqrt{1-\left(\frac{v^2}{c^2}\right)}} \quad \text{and} \quad x_1' = \frac{x_1 - vt}{\sqrt{1-\left(\frac{v^2}{c^2}\right)}}.$$

$$\therefore \quad L_0 = x_2' - x_1' = \frac{x_2 - x_1}{\sqrt{1-\left(\frac{v^2}{c^2}\right)}} = \frac{L}{\sqrt{1-\left(\frac{v^2}{c^2}\right)}}$$

or $$L = L_0\sqrt{1-\left(\frac{v^2}{c^2}\right)}.$$

The equation shows that to the stationary observer in S, the rod placed in the moving frame S' appears to be contracted by a factor $\sqrt{1-(v^2/c^2)}$. This contraction occurs in the direction of relative motion. If v is parallel to x, the v and z dimensions of the rod are same in both S and S'.

If v = c , then L = 0. It means that a rod moving with the speed of light will appear as reduced to a point to a stationary observer.

Proper Length : The length L_0 measured in the frame in which the rod is at rest is called 'proper' length.

(b) *Time Dilation : Time-intervals are also affected by relative motion. A clock moving with velocity v with respect to an observer appears him to have slowed down by a factor* $1/\sqrt{1-(v^2/c^2)}$, *than when at rest with respect to him.*

Suppose a clock is placed at the point x' in the moving frame S'. An observer in S' finds that the clock gives two ticks at times t_1' and t_2'. The time-interval between the ticks as judged from S' is

$$t_0 = t_2' - t_1'.$$

t_0 is the interval as measured in a frame in which the clock is at rest. Now, suppose another observer measures the time-interval (with his own clock) between the same two ticks from a stationary frame of reference S, relative to which the clock (or S') is moving with velocity v. If he records the ticks at times t_1 and t_2, the time-interval appears to him as

$$t = t_2 - t_1.$$

From the Lorentz transformation for a fixed value of x', we have

$$t_2 = \frac{t_2' + \left(\frac{x'v}{c^2}\right)}{\sqrt{1 = \left(\frac{v^2}{c^2}\right)}} \quad \text{and} \quad t_1 = \frac{t_1' + \left(\frac{x'v}{c^2}\right)}{\sqrt{1 - \left(\frac{v^2}{c^2}\right)}}$$

$$\therefore \qquad t = t_2 - t_1 = \frac{t_2' - t_1'}{\sqrt{1 - \left(\frac{v^2}{c^2}\right)}}$$

or

$$t = \frac{t_0}{\sqrt{1 - \left(\frac{v^2}{c^2}\right)}}.$$

This eq. shows that to the stationary observer in S, the time-interval appears to be *lengthened* by a factor $1/\sqrt{1 - (v^2/c^2)}$. Thus, a stationary clock measures a longer time-interval between events occurring in a moving frame of reference than does a clock in the moving frame. In other words, a moving clock appears to be slowed down to a stationary observer. This phenomenon is known as "time dilation".

If v = c, then t = ∞. It means that a clock moving with the speed of light will appear to be completely stopped to a stationary observer.

Proper Time-Interval : The proper time-interval is the time-interval measured by a clock attached to the observed body. In the above example, to is the proper time-interval. It is non-proper time-interval).

Experimental Verification of Time Dilation—μ-Meson Decay: Direct Experimental evidence for time-dilation has been found in cosmic ray Physics. Certain particles, called μ-mesons, are created by fast cosmic-ray particles at a height of about 10 kilometers from the surface of the earth and reach the earth in large numbers. They have a typical speed of 2.994×10^8 meters/sec, which is 0.998 of the speed of light c. Further, a μ-meson is found to have an average life-time of 2×10^{-6} second after which it decays into an electron. Thus, in its life- time a μ-meson can travel a distance of only $(2.994 \times 10^8 \text{ m/s}) \times (2 \times 10^{-6} \text{ s}) \simeq 600$ meters or 0.6 km. Then how do μ-mesons travel a distance of 10 km to reach the earth?

The answer is supplied by the special theory of relativity. The μ-meson has a life-time t_0 (= 2×10^{-6} second) *in its own frame of reference.* In the observer's frame of reference on the earth, however, the life-time is lengthened owing to the relative motion, to the value t given by

$$t = \frac{t_0}{\sqrt{1-\left(\frac{v^2}{c^2}\right)}} = \frac{2 \times 10^{-6}}{\sqrt{1-(0.998)^2}}$$

$$= \frac{2 \times 10^{-6}}{0.063} = 3.17 \times 10^{-5} \text{ second.}$$

In this life-time a meson whose speed is 0.998 c (= 2.994×10^8 meters/second) can travel a distance

$$(2.994 \times 10^8 \text{ m/s}) \times (3.17 \times 10^{-5} \text{ s})$$

$\simeq$ 9500 meters $\simeq$ 9.5 km.

Hence, despite their brief life-time, it is possible for the mesons to reach the earth from the large altitudes at which they are actually formed.

Example 22:

Show that two simultaneous events at different positions in a frame of reference are not in general simultaneous in another inertial frame in relative motion.

Hence show that there can be a disagreement about past and future between two observers in relative motion.

Solution:

Relative Character of Time—Concept of Simultaneity : An interesting implication of the relativity theory is that time-intervals are not the same for two observers in relative motion. This leads to an important fact that *two events which appear to fake place simultaneously to one observer are, in general, not simultaneous to another observer in relative motion.*

Suppose, two time-bombs explode at different places x_1 and x_2 but at the same time t_0 to an observer on the ground. The situation is different to a pilot of a spaceship moving with a velocity v relative to ground. To him, the explosion at x_1 occurs at

$$t_1' = \frac{t_0 - \left(\frac{x_1 v}{c^2}\right)}{\sqrt{1-\left(\frac{v^2}{c^2}\right)}}$$

and that at x_2 occurs at

$$t_2' = \frac{t_0 - \left(\frac{x_2 v}{c^2}\right)}{\sqrt{1-\left(\frac{v^2}{c^2}\right)}}.$$

Hence the two events (explosions) that occur simultaneously to one observer are separated to another observer by a time-interval of

$$t_2' - t_1' = \frac{(x_1 - x_2)\frac{v}{c^2}}{\sqrt{1-\left(\frac{v^2}{c^2}\right)}}.$$

Who is right ? The question is meaningless; both observers arc right because each one measures what he sees. Hence we conclude that there is no such thing as "absolute time" which is same for all observers. Time is relative and is different for observers in relative motion.

Past, Present and Future : The relative character of time, however, does not discard all of our everyday experiences regarding time. For example, no observer, regardless of his state of motion, can see an event before it actually occurs. This is because an event sends out light signal which takes finite time to reach the observer. Similarly, time never runs backward to a/iv observer. A sequence of events that occur somewhere at t_1, t_2 ,t_3 ...... will appear in the same order to all observers everywhere, though not necessarily with the same time-intervals $t_2 - t_1$, $t_3 - t_2$ .... between each pair of events. Thus, we can order events in time. Hence there is a sense of lime distinguishing past, present and future.

However, it is possible that an observer has seen an event occurring at a distant place, whereas a second farther observer has yet to sec it. Thus, the event has occurred in the past of the first observer, but has to occur in the future of the second observer. (In other words, the two observers disagree about past and Future.) Can the first observer send a radio message about this event to the second observer enabling him to predict the future? The answer is "no" because the radio message has to travel through space with the speed of light and can never reach the second observer before he himself becomes aware of the event. Thus, even in the theory of relativity, there is no way to peep into the future.

Example 23:

The total energy of a particle is exactly twice its rest energy. Find its speed.

Solution:

The total energy E of a moving particle is mc^2 while its rest energy E_0 is m_0c^2. When $E = 2E_0$, then

$$mc^2 = 2m_0c^2$$

or $$m = 2m_0.$$

The mass of the moving particle is related to its rest m_0 by

$$m = \frac{m_0}{\sqrt{1-\left(\frac{v}{c}\right)^2}}.$$

Here $m = 2m_0$.

$$\therefore \quad 2m_0 = \frac{m_0}{\sqrt{1-\left(\frac{v}{c}\right)^2}}$$

or $$1-\frac{v^2}{c^2} = \frac{1}{4}$$

or $$v = \frac{\sqrt{3}}{2}c = 0.886\ c.$$

This result does *not* depend upon the rest mass of the particle.

Example 24:

What is the speed of a 2.0 MeV electron, where $m_0c^2 = 0.51$ MeV?

Solution:

The energy of the electron is given by

$$E = \frac{m_0c^2}{\sqrt{1-\left(\frac{v}{c}\right)^2}}.$$

Here E = 2.0 MeV and m_0c^2 = 0.51 MeV.

$$\therefore \qquad \sqrt{1-\left(\frac{v}{c}\right)^2} = \frac{m_0c^2}{E} = \frac{0.51\,\text{MeV}}{2.0\,\text{MeV}} = 0.255$$

$$\text{or} \qquad 1-\frac{v^2}{c^2} = (0.255)^2 = 0.065$$

$$\text{or} \qquad \frac{v^2}{c^2} = 1 - 0.065 = 0.935$$

$$\text{or} \qquad v = \sqrt{0.935}\,c = 0.967\ c.$$

Example 25:

A body of mass m_0 at rest break up spontaneously into two parts, having rest masses m_{01} and m_{02} and speeds v_1 and v_2 respectively. Show that $m_0 > m_{01} + m_{02}$.

Solution:

The total energy of the body at rest is its rest mass energy, m_0c^2. It breaks up into two moving parts. The energies (kinetic + rest mass) of the parts are $\frac{m_{01}c^2}{\sqrt{1-(v_1/c)^2}}$ and $\frac{m_{02}c^2}{\sqrt{1-(v_2/c)^2}}$. By conservation of mass-energy, we have

$$m_0c^2 = \frac{m_{01}c^2}{\sqrt{1-\left(\frac{v_1}{c}\right)^2}} + \frac{m_{02}c^2}{\sqrt{1-\left(\frac{v_2}{c}\right)^2}}.$$

Both $\sqrt{1-\left(\frac{v_1}{c}\right)^2}$ and $\sqrt{1-\left(\frac{v_2}{c}\right)^2}$ are less than 1.

$$\therefore \qquad m_0 > m_{01} + m_{02}.$$

Example 26:

Calculate the length and the orientation of a rod of length 5 m in a frame of reference which is moving with a velocity equal to 0.6 c, in a direction making an angle of 30° with the rod.

Solution:

The component of the length of the 5-meter rod along the direction of motion of the frame is L_{x0} = 5 cos 30° meter and that perpendicular to this direction is L_{y0} = 5 sin 30° meter.

The apparent length along the moving frame is

$$L_x = L_{x0}\sqrt{1-\frac{v^2}{c^2}}$$

$$= (5\cos 30^\circ \text{ m})\sqrt{1-(0.6)^2}$$

$$= (5 \cos 30^\circ \text{ m})\ 0.8 = 2\sqrt{3} \text{ meter.}$$

The length perpendicular to the direction of motion remains the same, that is,

$$L_y = L_{y0} = 5 \sin 30^\circ = (5/2) \text{ meter.}$$

∴ length observed in the moving frame

$$L = \sqrt{L_x^2 + L_y^2} = \sqrt{12+\frac{25}{4}} = \sqrt{18.25} = 4.272 \text{ meter.}$$

If the rod appears to make an angle θ with the direction of motion, then

$$\tan\theta = \frac{L_y}{L_x} = \frac{5/2}{2\sqrt{3}} = \frac{5}{4\sqrt{3}} = 0.72.$$

$$\therefore \theta = \tan^{-1}(0.72) = 36^\circ.$$

Example 27:

What is the fractional difference in mass between a 1.0 gram piece of copper at 0°C and the same piece of copper at 100°C? The specific heat of copper is 0.093 cal/g-°C. (1 cal = 4.18 J)

Solution:

The heat energy absorbed by the copper piece is

$$\Delta E = \text{mass} \times \text{sp. heat} \times \text{temp-rise}$$

$$= 1.0\text{g} \times \frac{0.093\text{ cal}}{\text{g–°C}} \times 100°\text{C}$$

$$= 9.3 \text{ cal} = 9.3 \times 4.18 = 38.87 \text{ J.}$$

∴ increase in mass is given by

$$\Delta m = \frac{\Delta E}{c^2} = \frac{38.87\text{J}}{\left(3 \times 10^8 \text{ m/s}\right)^2} = 4.32 \times 10^{-16} \text{ kg}$$

$$= 4.32 \times 10^{-13} \text{ g.}$$

Fractional increase is

$$\frac{\Delta m}{m} = \frac{4.32 \times 10^{-13}\text{g}}{1.0\text{g}} = 4.32 \times 10^{-13}.$$

Example 28:

How much mass is lost when 1 kg of water at 0°C turns of ice at 0°C? (Latent heat of ice is 80 cal/g and 1 cal = 4.18 × 10^7 erg; c = 3 × 10^{10} cm/s).

Solution:

The heat (energy) withdrawn from the water to freeze it at 0°C is

$$\Delta E = mL$$

$$= 1000 \text{ g} \times 80 \text{ cal/g} = 8 \times 10^4 \text{ cal}$$

$$= (8 \times 10^4) \times (4.18 \times 10^7) = 3.344 \times 10^{12} \text{ erg.}$$

Let Δm be the corresponding loss in mass. From Einstein's mass-energy equivalence, we have

$$\therefore \quad \Delta m = \frac{\Delta E}{c^2} = \frac{3.344 \times 10^{12} \text{erg}}{\left(3 \times 10^{10} \text{cm/s}\right)^2}$$

$$= 3.72 \times 10^{-9} \text{ g.}$$

Example 29:

(a) Explain why we cannot accelerate an electron to a velocity greater than the velocity of light in free space. Can it be so in a material medium?

Or

Electron beam in a television picture tube can move across the screen at a speed faster than the speed of light. Does this violate the special theory of relativity?

(b) Define momentum and force relativistically and show that the rest mass of a photon is zero.

Solution:

(a) *Limiting Velocity of a Material Particle* : The mass of a body moving relative to an observer increases with increasing velocity according to the equation

$$m = \frac{m_0}{\sqrt{1-\left(\frac{v^2}{c^2}\right)}},$$

where m is the mass at velocity v and m_0 is the rest mass. Thus, as the body's velocity v approaches the velocity of light c, the ratio $v^2/c^2 \to 1$ and the quantity $\sqrt{1-(v^2/c^2)} \to 0$. Hence the mass m of the body approaches infinity, as v approaches c . For instance, an electron energised to 10^9 eV has an effective mass nearly 2000 times its rest mass.

The variation of mass with velocity, therefore, places a limit on the velocity which a material particle can have relative to an observer. As we accelerate the particle, its mass becomes larger and larger and it becomes increasingly more difficult to accelerate it further, Infact, since the mass would become infinite when v = c , an infinite force would be needed to accelerate the particle to the velocity of light c. Since no infinite forces are in the universe, the velocity of a particle can never be made to exceed the velocity of light in free space. Although it is true 'that a moving particle can never travel faster than the velocity of light *in free space* (c = 3 × 10^8 m/s), this is not true in a material medium (water, glass, etc.). In all material media light travels more slowly than its velocity in free space. Very energetic atomic panicles are, however, capable of moving faster in a material medium than the velocity of light *in that medium,* though never faster than the velocity of light in free space. This phenomenon gives rise to light waves known as, 'cerenkov radiation.' Hence the motion of high-energy electron beam *across the screen* faster than light does not violate the theory of relativity.

(b) *Relativistic Momentum and Force* : In relativity, the momentum p of a particle at any velocity v can be defined in the same way as in classical mechanics, that is,

$$p = mv.$$

But $m = m_0/\sqrt{1-(v^2/c^2)}$, where my is the rest mass of the particle.

$$\therefore \quad p = \frac{m_0 v}{\sqrt{1-\left(\frac{v^2}{c^2}\right)}}.$$

With this definition of momentum, the law of conservation of momentum is true in special relativity just as in classical Physics.

In relativity, however, the force acting on a body *cannot* be defined as mass multiplied by acceleration of the body. It must be defined only as rate of change of momentum (Newton's original definition). Thus

$$F = \frac{dp}{dt} = \frac{d}{dt}\left[\frac{m_0 v}{\sqrt{1-\left(\frac{v^2}{c^2}\right)}}\right].$$

Rest Mass of Photon : A photon travels with the velocity of light, and hence we must use relativistic expression for its momentum p. Thus, for a photon

$$p = \frac{m_0 v}{\sqrt{1-\left(\frac{v^2}{c^2}\right)}},$$

where m_0 is the rest mass and v the velocity of the photon.

From quantum theory of radiation, the momentum of a photon of radiation of wavelength λ is $p = h/\lambda$, where h is the Planck's constant. Therefore

$$\frac{h}{\lambda} = \frac{m_0 v}{\sqrt{1-\left(\frac{v^2}{c^2}\right)}}.$$

or
$$m_0 = \frac{h}{v\lambda}\sqrt{1-\left(\frac{v^2}{c^2}\right)}.$$

Since for photon, $v = c$, we have

$$m_0 = 0.$$

Hence the rest mass of photon is zero.

Example 30:

(a) Deduce Einstein's mass-energy relation $E = mc^2$ and discuss it. Give some evidence showing its validity.

Or

Show that particle's total energy is $E = mc^2$.

Or

Show that kinetic energy $K = (m - m_0) c^2$.

(b) A spring is compressed by tying its ends together and then completely dissolved in acid. Where did its stored potential energy go?

Solution:

(a) *Einstein's Mass-Energy Equivalence Relation* : According to Einstein's special theory of relativity, the mass m of a body moving with velocity v relative to a stationary observer varies with v and is given by

$$m = \frac{m_0}{\sqrt{1 - \left(\frac{v^2}{c^2}\right)}}, \quad \text{...(i)}$$

where m_0 is the 'rest mass' of the body and c the velocity of light.

The variation of mass with velocity has modified our ideas about energy. Let us consider a particle of mass m acted upon by a force F in the same direction as its velocity v . The force is defined as the rate of change of momentum *i.e.,*

$$F = \frac{d}{dt}(mv)$$

$$= m\frac{dv}{dt} + v\frac{dm}{dt}. \quad \text{(m is variable)}$$

The work done by the force F in a displacement ds of the body is equal to the change in the kinetic energy K of the body. Thus

$$dK = F ds = m\frac{dv}{dt}ds + v\frac{dm}{dt}ds$$

$$= m\frac{ds}{dt}dv + v\frac{ds}{dt}dm$$

$= mv\ dv + v^2\ dm,$ ...(ii)

because $ds/dt = v$.

Now, let us differentiate eq. (i):

$$dm = m_0\left(-\frac{1}{2}\right)\left(1-\frac{v^2}{c^2}\right)^{-3/2}\left(-\frac{2v\,dv}{c^2}\right)$$

$$= \frac{m_0}{c^2}\frac{v\,dv}{\left(1-\frac{v^2}{c^2}\right)^{3/2}}.$$

But $$m_0 = m\left(1-\frac{v^2}{c^2}\right)^{1/2}$$ by eq. (i).

$$\therefore \qquad dm = \frac{mv\,dv}{\left(c^2 - v^2\right)}$$

or $mv\ dv = (c^2 - v^2)\ dm$.

Making this substitution in eq. (ii), we get

$$dK = (c^2 - v^2)\ dm + v^2\ dm = c^2\ dm.$$

Suppose, the body has a mass my when at rest and a mass m when accelerated to a velocity v. The kinetic energy acquired is then

$$K = \int dK = \int_{m_0}^{m} c^2\, dm$$

or $\qquad K = (m - m_0)\ c^2.$ ...(iii)

This result shows that the kinetic energy of a body is equal to the relativistic increase in mass of the body over the rest mass multiplied by the square of the velocity of light. It can be interpreted to mean that even when the body is at rest, it possesses an amount of energy $m_0\ c^2$. Accordingly, $m_0\ c^2$ is called the 'rest energy' E_0 of a body whose rest mass is m_0. The *total* energy E of the body is therefore the sum of the kinetic energy and the rest energy. Thus

$$E = K + E_0 = (m - m_0)\ c^2 + m_0c^2$$

or $$E = mc^2.$$

This is the famous Einstein's mass-energy relation. This shows that with the mass m of a system is associated an amount of energy me or, conversely, a system with total energy E has associated with it an inertial mass E/c^2.

Whenever we supply any kind of energy to a body, the mass of the body increases. In fact, a body when heated becomes more massive.

Verification : The mass and energy relation has been verified in a number of phenomena:

(i) *Compton Effect :* Compton treated the X-ray scattering as an elastic collision between a photon and an electron, after which the scattered photon moves in a new direction and the electron recoils with a velocity comparable with the velocity of light. Applying the laws of conservation of energy and momentum in the light of mass-energy relation, Compton calculated a value for the wavelength shift which agreed with the experimental results.

(ii) *Fine Structure of Spectral Lines :* Sommerfeld explained the fine structure of spectral lines on the basis of relativistic variation of mass. The nice agreement of his theory with experiment provides another verification of mass-energy relation.

(iii) *Nuclear Phenomena :* The explanation of 'mass defect' and 'release of tremendous amount of energy in nuclear fission' which is entirely based on mass-energy relation, gives a strong support to the relation.

(b) *Compressed Spring dissolved in Acid :* The idea of mass-energy equivalence can be extended to energies other than kinetic. For example, when we compress a spring, we give to it elastic potential energy ΔU (say) and its mass increases by an amount $\Delta U/c^2$. Thus, the mass of a compressed spring is greater than that of an uncompressed one. If the compressed spring is dissolved in acid, the reaction products are of slightly greater mass (though undetectable) than if the spring were uncompressed.

Example 31:

A rod 1.0 meter long is moving along its length with a velocity of 0.6 c. Calculate the length as it appears to a stationary observer.

Solution:

As a consequence of Lorentz transformation equations, when a body moves with a velocity v relative to the observer, its measured length is contracted in the direction of its motion by the factor $\sqrt{1-(v^2/c^2)}$ Thus, if L_0 be the true length (when at rest), the length when in motion is

$$L = L_0\sqrt{1-\left(\frac{v^2}{c^2}\right)}$$

When v = 0.6 c, then the apparent length of the rod (L_0 = 1.0 m) would be

$$L = L_0\sqrt{1-(0.6)^2} = 0.8\,L_0$$
$$0.8 \times 1.0\,m = 0.8\,m.$$

Example 32:

Using the concept of variation of mass with velocity, derive relativistic expression for the kinetic energy of a particle. Show that for small velocities the relativistic kinetic energy reduces to the classical kinetic energy. If a particle could move with the velocity of light how much kinetic energy it would possess?

Solution:

Relativistic Kinetic Energy of a Body : We have deduced in the last question that the relativistic kinetic energy of a particle of rest mass my is given by

$$K = (m - m_0)c^2.$$

But $m = \dfrac{m_0}{\sqrt{1-(v^2/c^2)}}$, where in is the mass of the particle moving with velocity v. Therefore

$$K = \left[\frac{m_0}{\sqrt{1-\left(\frac{v^2}{c^2}\right)}} - m_0\right]c^2$$

or

$$K = m_0c^2\left[\left(1-\frac{v^2}{c^2}\right)^{-1/2} - 1\right]. \qquad ...(i)$$

This is the relativistic kinetic-energy expression in terms of velocity v.

Expanding $\left(1-\frac{v^2}{c^2}\right)^{-1/2}$ by binomial theorem, we get

$$\left(1-\frac{v^2}{c^2}\right)^{-1/2} = 1+\frac{1}{2}\frac{v^2}{c^2}+\frac{3}{8}\frac{v^4}{c^4}+\ldots$$

For small velocities v << c, the terms beyond $\frac{1}{2}\frac{v^2}{c^2}$ are negligible.

Then $$\left(1-\frac{v^2}{c^2}\right)^{-1/2} \approx 1+\frac{1}{2}\frac{v^2}{c^2}.$$

Making this substitution in eq. (i), we obtain

$$K \simeq m_0c^2\left\{1+\frac{1}{2}\frac{v^2}{c^2}-1\right\} = \frac{1}{2}m_0v^2,$$

which is the classical expression for the kinetic energy. The classical kinetic energy is less than the rest energy (m_0c^2).

We have seen above that the relativistic kinetic energy is

$$K = \left[\frac{m_0}{\sqrt{1-\left(\frac{v^2}{c^2}\right)}} - m_0\right]c^2.$$

If v = c, then $\sqrt{1-(v^2/c^2)} = 0$ and so

$$K = \infty.$$

Example 33:

Derive the formula $E^2 = m_0^2c^4 + p^2c^2$,

where symbols have their usual meanings.

Hence obtain relativistic relation between kinetic energy and momentum.

Solution:

Relativistic Relation between Energy and Momentum : The total (rest-mass plus kinetic) energy of a particle is given by

$$E = mc^2 = \frac{m_0c^2}{\sqrt{1-\left(\frac{v^2}{c^2}\right)}}, \qquad \text{...(i)}$$

and the relativistic momentum is given by

$$p = mv = \frac{m_0 v}{\sqrt{1-\left(\frac{v^2}{c^2}\right)}} \qquad \text{...(ii)}$$

and so

$$pc = \frac{m_0 vc}{\sqrt{1-\left(\frac{v^2}{c^2}\right)}}. \qquad \text{...(iii)}$$

Squaring and subtracting eq. (iii) from eq. (i), we get

$$E^2 - p^2c^2 = \frac{m_0^2c^4 - m_0^2v^2c^2}{1-\frac{v^2}{c^2}}$$

$$= \frac{m_0^2c^4\left(1-\frac{v^2}{c^2}\right)}{1-\frac{v^2}{c^2}} = m_0^2c^4$$

or $\quad E^2 = m_0^2c^4 + p^2c^2$

or $\quad E = \sqrt{m_0^2c^4 + p^2c^2}$.

This is the relativistic relation between total energy E and momentum p of a particle.

Relativistic Relation between Kinetic Energy and Momentum : The relativistic kinetic energy of a particle is total energy minus rest-mass energy, that is,

$$K = E - m_0c^2$$

$$= \sqrt{m_0^2c^4 - p^2c^2} - m_0c^2$$

$$= m_0c^2\left\{\left(1+\frac{p^2}{m_0^2c^2}\right)^{1/2} - 1\right\}.$$

This is the required relation.

When v << c, *i.e.*, p << m_0c, then we can write

$$K \simeq m_0c^2 \left\{\left(1+\frac{1}{2}\frac{p^2}{m_0^2c^2}\right)-1\right\}$$

or $$K \simeq \frac{p^2}{2m_0}.$$

Again, when v << c, then m

∴ $$K \simeq \frac{p^2}{2m}.$$

Thus, in the limit of small velocities, the relativistic relation between kinetic energy and momentum tends to the classical relation.

Example 34:

(a) What are massless particles?

Show that massless particles can exist only if they move with the speed of light and that their energy E and momentum p must have the relation E = pc.

(b) If photons have a speed c in one reference frame, they can be found at rest in any other frame?

Solution:

(a) *Massless Particles : A massless particle is one which has a zero rest mass.* Classically, such a particle cannot exist. In relativistic mechanics, however, a particle having zero rest mass can exist and show particle-like properties as energy and momentum.

Let us see the requirement for the existence of a massless particle. The relativistic total energy and momentum of a particle of rest mass my moving with speed v are given by

$$E = mc^2 = \frac{m_0c^2}{\sqrt{1-\left(\frac{v^2}{c^2}\right)}}$$

and $$p = mv = \frac{m_0v}{\sqrt{1-\left(\frac{v^2}{c^2}\right)}}.$$

When $m_0 = 0$ and $v < c$, then $E = p = 0$. A massless particle with a speed less than c can have neither energy nor momentum. However, if $m_0 = 0$ and $v = c$, both E and p become indeterminate and can have any values. Hence *massless particles having energy and momentum can exist provided that they travel with the speed of light.* Photon is an example of massless particle.

Again, the relativistic total energy E of a particle of rest mass my in terms of its momentum p is given by

$$E = \sqrt{m_0^2 c^4 + p^2 c^2} .$$

If $m_0 = 0$, then

$$E = pc.$$

This is also the result obtained in classical electromagnetism for radiation (or photon whose rest mass is zero).

(b) *Speed of Photon* : When a body moves with a speed less than c in one reference frame, we can find another reference frame in which it is at rest. But when it moves with speed c in one reference frame, it will move with the same speed c in all reference frames. Since-the photon has a zero rest mass it should always move with the speed of light, whatever is the reference frame.

Example 35:

Calculate the speed of a particle of rest mass 3.33×10^{-27} gram whose energy is 2 MeV.

Solution:

The (total) energy of a particle of rest mass m_0 and moving with a speed v is given by

$$E = mc^2 = \frac{m_0 c^2}{\sqrt{1 - \left(\frac{v}{c}\right)^2}} .$$

Here $\quad E = 2 \text{ MeV} = 2 \times (1.6 \times 10^{-13})$ joule

and $\quad m_0 = 3.33 \times 10^{-30}$ kg.

$$\therefore \quad 2 \times 1.6 \times 10^{-13} = \frac{\left(3.33 \times 10^{-30}\right)\left(3.0 \times 10^{8}\right)^2}{\sqrt{1 - \left(\frac{v}{c}\right)^2}}$$

or $$\sqrt{1-\left(\frac{v}{c}\right)^2} = \frac{30 \times 10^{-14}}{2 \times 1.6 \times 10^{-13}} = 0.937$$

or $$1-\frac{v^2}{c^2} = (0.937)^2 = 0.878$$

or $$\frac{v^2}{c^2} = 1 - 0.878 = 0.122.$$

or $$\frac{v}{c} = \sqrt{0.122} = 0.349$$

or $$v = 0.349\ c = 0.349 \times (3 \times 10^8\ \text{m/s})$$

$$= 1.047 \times 10^8\ \text{m/s}.$$

Example 36:

Find expressions for the energy and momentum of a particle measured in a moving reference frame on the basis of Lorentz transformation.

Solution:

Transformation of Energy and Momentum : Let there be two reference frames S and S', the frame S' moving with a constant velocity v relative to S. Let u and u' be the velocities of a particle measured by observers in S and S' respectively. By the relativistic addition of velocities, we have

$$u' = \frac{u - v}{1-\left(\frac{uv}{c^2}\right)} \qquad ...(i)$$

The energy and momentum of the particle in the frame S are

$$E = mc^2 = \frac{m_0c^2}{\sqrt{1-\left(\frac{u^2}{c^2}\right)}} \qquad ...(ii)$$

and $$p = mu = \frac{m_0u}{\sqrt{1-\left(\frac{u^2}{c^2}\right)}}. \qquad ...(iii)$$

The corresponding quantities in the frame S' are

$$E' = m'c^2 = \frac{m_0 c^2}{\sqrt{1-\left(\frac{u'^2}{c^2}\right)}} \quad \text{...(iv)}$$

and $$p' = m'u = \frac{m_0 u'}{\sqrt{1-\left(\frac{u'^2}{c^2}\right)}}. \quad \text{...(v)}$$

Now, $$\left[1-\frac{u'^2}{c^2}\right]^{-1/2} = \left[1-\frac{1}{c^2}\frac{(u-v)^2}{\left(1-\frac{uv}{c^2}\right)}\right]^{-1/2}$$

$$= \left[\frac{\left(1-\frac{uv}{c^2}\right)^2 - \frac{(u-v)^2}{c^2}}{\left(1-\frac{uv}{c^2}\right)^2}\right]$$

$$= \left[\frac{1+\frac{u^2v^2}{c^4}-\frac{2uv}{c^2}-\frac{u^2}{c^2}-\frac{v^2}{c^2}+\frac{2uv}{c^2}}{\left(1-\frac{uv}{c^2}\right)^2}\right]^{-1/2}$$

$$= \left[\frac{\left(1-\frac{u^2}{c^2}\right)-\frac{v^2}{c^2}\left(1-\frac{u^2}{c^2}\right)}{\left(1-\frac{uv}{c^2}\right)^2}\right]^{-1/2}$$

$$= \left[\frac{\left(1-\frac{u^2}{c^2}\right)\left(1-\frac{v^2}{c^2}\right)}{\left(1-\frac{uv}{c^2}\right)^2}\right]^{-1/2}$$

$$= \frac{1-\frac{uv}{c^2}}{\left(1-\frac{u^2}{c^2}\right)^{1/2}\left(1-\frac{v^2}{c^2}\right)^{1/2}} \quad \text{...(vi)}$$

We make this substitution in eq. (iv).

$$E' = m_0c^2 \frac{\left(1-\frac{uv}{c^2}\right)}{\left(1-\frac{u^2}{c^2}\right)^{1/2}\left(1-\frac{v^2}{c^2}\right)^{1/2}}$$

$$= \frac{1}{\left(1-\frac{v^2}{c^2}\right)^{1/2}}\left[\frac{m_0c^2}{\left(1-\frac{u^2}{c^2}\right)^{1/2}} - \frac{m_0\,uv}{\left(1-\frac{u^2}{c^2}\right)^{1/2}}\right].$$

Using eq. (ii) and (iii), we get

$$E' = \frac{E - pv}{\left(1-\frac{v^2}{c^2}\right)^{1/2}}. \qquad ...(viii)$$

Again, making substitutions from eq.(vi) and eq.(i) in eq.(v), we get

$$p' = m_0 \frac{u-v}{1-\frac{uv}{c^2}} \frac{1-\frac{uv}{c^2}}{\left(1-\frac{u^2}{c^2}\right)^{1/2}\left(1-\frac{v^2}{c^2}\right)^{1/2}}$$

$$= \frac{1}{\left(1-\frac{v^2}{c^2}\right)^{1/2}}\left[\frac{m_0u}{\left(1-\frac{u^2}{c^2}\right)^{1/2}} - \frac{m_0v}{\left(1-\frac{u^2}{c^2}\right)^{1/2}}\right].$$

Using eq. (iii) and (ii), we get

$$p' = \frac{p - \frac{Ev}{c^2}}{\left(1-\frac{v^2}{c^2}\right)^{1/2}}. \qquad ...(viii)$$

Eq. (vii) and (viii) are the required expressions.

Example 37:

Prove that $E^2 - p^2c^2 = m_0^2c^4$ is invarient under Lorentz transformation.

Solution:

The right-hand side $m_0^2c^4$ is obviously invariant. Hence it is enough to show that

$$E'^2 - p'^2c^2 = E^2 - p^2c^2,$$

where E' and p' are the energy and momentum in a frame S' moving uniformly with a velocity v relative to S in which the energy and momentum are E and p.

The Lorentz transformation equations for energy and momentum are

and
$$p' = \frac{p - \dfrac{Ev}{c^2}}{\left(1 - \dfrac{v^2}{c^2}\right)^{1/2}}.$$

Now, $$E'^2 - p'^2c^2 = \frac{\left(E^2 + p^2v^2 - 2\,Epv\right) - \left(p^2 + \dfrac{E^2v^2}{c^4} - \dfrac{2pEv}{c^2}\right)c^2}{1 - \dfrac{v^2}{c^2}}$$

$$= \frac{E^2 + p^2v^2 - 2Epv - p^2c^2 - \dfrac{E^2v^2}{c^2} + 2pEv}{1 - \dfrac{v^2}{c^2}}$$

$$= \frac{E^2\left(1 - \dfrac{v^2}{c^2}\right) - p^2c^2\left(1 - \dfrac{v^2}{c^2}\right)}{1 - \dfrac{v^2}{c^2}} = E^2 - p^2c^2$$

Thus, $E^2 - p^2c^2$ is Lorentz invariant.

Example 38(a):

A charged particle shows an acceleration of 4.2×10^{12} cm/s² under an electric field at low speed. Compute the acceleration of the particle under the same field when the speed has reached a value 2.88×10^{10} cm/s. The speed of light is 300×10^{19} cm/s.

Solution:

The force F on a particle having charge q under an electric field

$$F = qE.$$

At low speed, the mass of the particle is equal to its rest mass m_0. Therefore, the acceleration at low speed is

$$a_0 = \frac{F}{m_0} = \frac{qE}{m_0} = 4.2 \times 10^{12}\ \text{cm/s}^2.$$

When the particle attains a speed 2.88×10^{10} cm/s which is comparable to the speed of light, its mass increases to m, where

$$m = \frac{m_0}{\sqrt{1-\left(\frac{v}{c}\right)^2}} = \frac{m_0}{\sqrt{1-\left(\frac{2.88 \times 10^{10}}{3.00 \times 10^{10}}\right)^2}} = \frac{m_0}{0.28}.$$

Now, the acceleration would be

$$a = \frac{F}{m} = 0.28\frac{F}{m_0}$$

$$= 0.28 \times (4.2 \times 10^{12}) = 1.176 \times 10^{12}\ \text{cm/s}^2.$$

Example 38(b):

The rest mass of an electron is 9 × 10 ³¹ kg. What will be its mass if it were moving with 4/5th the speed of light?

Solution:

The mass of a body (rest mass my) moving with velocity v is given by

$$m = \frac{m_0}{\sqrt{1-\left(\frac{v}{c}\right)^2}}.$$

Here $\frac{v}{c} = \frac{4}{5}$, and so $\sqrt{1-\left(\frac{v}{c}\right)^2} = \sqrt{1-\left(\frac{16}{25}\right)} = \frac{3}{5}$.

$$\therefore \qquad m = \frac{9 \times 10^{-31}\ \text{kg}}{3/5} = 15 \times 10^{-31}\ \text{kg}.$$

Example 38(c):

Deduce the velocity at which the mass of a particle becomes 1.25 times its rest mass. (c = 3 × 10⁸ m/s).

Solution:

$$m = \frac{m_0}{\sqrt{\left(\frac{v}{c}\right)^2}}.$$

For $\frac{m}{m_0} = 1.25$, we have

$$\frac{1}{\sqrt{1-\left(\frac{v}{c}\right)^2}} = 1.25 = \frac{5}{4}$$

or
$$1 - \frac{v^2}{c^2} = \frac{16}{25}$$

or
$$\frac{v^2}{c^2} = 1 - \frac{16}{25} = \frac{9}{25}.$$

This gives $v = 0.6\ c = 1.8 \times 10^8$ meter/second.

Example 39:

Deduce the rest energy of an electron in joules and in electron-volts ($m_0 = 9.1 \times 10^{-31}$ kg, $c = 3.0 \times 10^8$ m/s). Also deduce the speed at which the total relativistic energy becomes 1.25 times the rest energy.

Solution:

The rest energy is

$$E_0 = m_0c^2$$
$$= (9.1 \times 10^{-31}\ \text{kg}) \times (3.0 \times 10^8\ \text{m/s})^2$$
$$= 8.19 \times 10^{-14}\ \text{joule.}$$

The 1 electron = volt = 1.6×10^{-19} joule.

$$\therefore \qquad E_0 = \frac{8.19 \times 10^{-14}}{1.6 \times 10^{-19}} = 0.51 \times 10^6\ \text{eV} = 0.51\ \text{MeV}.$$

The ratio of the total relativistic energy E to the rest energy E_0 is given as 1.25. That is,

$$\frac{E}{E_0} = \frac{mc^2}{m_0c^2} = \frac{m}{m_0} = 1.25.$$

But $$\frac{m}{m_0} = \frac{1}{\sqrt{1-\left(\frac{v}{c}\right)^2}}$$

$$\therefore \quad \frac{1}{1-\left(\frac{v^2}{c^2}\right)} = (1.25)^2 = 1.5625$$

or $$1-\frac{v^2}{c^2} = \frac{1}{1.5625}$$

or $$\frac{v^2}{c^2} = 1-\frac{1}{1.5625} = \frac{0.5625}{1.5625}.$$

$$\therefore \quad v = \sqrt{\frac{0.5625}{1.5625}}\,c = 0.6c.$$

Example 40:

Show that the mass of an electron is equivalent to 0.51 MeV energy. State the minimum energy of γ-ray photon which can produce an electron-positron pair.

Solution:

The rest mass m_0 of electron is 9.1×10^{-31} kg. As shown in the last problem, its energy equivalent is

$$E_0 = 0.51 \text{ MeV}.$$

To produce an electron-positron pair, the minimum energy of the γ-ray photon is the sum of the rest-mass energies of an electron and a positron, that is,

$$0.51 + 0.51 = 1.02 \text{ MeV}.$$

Example 41:

Obtain proton rest energy in MeV.

Solution.

$$\begin{aligned} E_0 &= m_0c^2 \\ &= (1.67 \times 10^{-27}\text{ kg})\,(3.0 \times 10^8\text{ m/s})^2 \\ &= 1.5 \times 10^{-10}\text{ J} \end{aligned}$$

$$= \frac{1.5\times10^{-10}}{1.6\times10^{-13}} = 937.5 \text{ MeV}.$$

Example 42(a):

Tell whether the following statement is true or false, giving reason. The momentum of an electron accelerated by a potential difference of 10^6 volt is 5.36×10^{-22} kg-m-s^{-1}, given rest mass of electron $m_0 = 9.1 \times 10^{-31}$ kg and $e = 1.6 \times 10^{-19}$ C.

Solution:

The statement is *false*.

If the momentum m_0v is 5.36×10^{-22} kg-m-s^{-1}, then the velocity of the electron would be

$$v = \frac{5.36\times10^{-22}\,\text{kg}-\text{m}-\text{s}^{-1}}{9.1\times10^{-31}\,\text{kg}} = 5.9 \times 10^{8} \text{ m s}^{-1},$$

which is greater than the velocity of light and hence impossible.

Example 42(b):

A particle of rest mass m_0 moves with speed $c/\sqrt{2}$. Calculate its mass, momentum, total energy and kinetic energy.

Solution:

The relativistic mass of the particle is

$$m = \frac{m_0}{\sqrt{1-\left(\frac{v}{c}\right)^2}} = \frac{m_0}{\sqrt{\left(1-\frac{1}{2}\right)}} = \sqrt{2}\,m_0$$

$$= 1.41\ m_0.$$

The momentum is given by

$$p = mv = \sqrt{2}m_0 \times \frac{c}{\sqrt{2}} = m_0c.$$

The total energy is

$$E = mc^2 = 1.41\ m_0c^2.$$

The kinetic energy is

$$K = E - m_0c^2 = 0.41\ m_0c^2.$$

Example 42(c):

A particle of rest mass m_0 initially moves with velocity 0.40 c . If the particle's velocity be doubled, how will its new momentum be compared with the initial momentum?

Solution:

The relativistic momentum is given by

$$p = \frac{m_0 v}{\sqrt{1-\left(\frac{v}{c}\right)^2}}.$$

Let p_i and p_f be the momenta when v = 0.40 c and v = 0.80 c respectively. Then

$$p_i = \frac{m_0(0.40c)}{\sqrt{1-(0.40)^2}} = \frac{m_0(0.40c)}{0.92}$$

and $$p_f = \frac{m_0(0.80c)}{\sqrt{1-(0.80)^2}} = \frac{m_0(0.80c)}{0.60}.$$

$$\therefore \qquad \frac{p_f}{p_i} = \frac{0.80}{0.60} \times \frac{0.92}{0.40} = 3.1.$$

Example 43:

What is the kinetic energy of a proton (rest mass 1.67×10^{-27} kg) moving with velocity 2.7×10^8 m/s. Express the result in MeV. ($c = 3.0 \times 10^8$ m/s).

Solution:

The velocity of the proton is comparable to the velocity of light. The (relativistic) kinetic energy is

$$K = m_0c^2 \left\{ \frac{1}{\sqrt{1-\left(\frac{v}{c}\right)^2}} - 1 \right\}.$$

Here $\frac{v}{c} = \frac{2.7 \times 10^8 \text{ m/s}}{3.0 \times 10^8 \text{ m/s}} = 0.9$. Thus

$$\sqrt{1-\left(\frac{v}{c}\right)^2} = \sqrt{1-(0.9)^2} = \sqrt{0.19} = 0.436.$$

$\therefore$ $$K = (1.67 \times 10^{-27} \text{ kg})(3.0 \times 10^8 \text{ m/s}^2)\left\{\frac{1}{0.436} - 1\right\}$$

$$= 1.94 \times 10^{-10} \text{ J}.$$

But 1 MeV = 1.60×10^{-13} J.

$\therefore$ $$K = \frac{1.94 \times 10^{-10}}{1.60 \times 10^{-13}} = 1212 \text{ MeV}.$$

Example 44:

How much work must be done in order to increase the speed of an electron from 1.8×10^8 m/s to 2.4×10^8 m/s.

Solution:

The initial (relativistic) kinetic energy of the electron (rest mass $m_0 = 9.1 \times 10^{-31}$ kg) is

$$K_1 = m_0c^2\left\{\left(1 - \frac{v^2}{c^2}\right)^{-1/2} - 1\right\}$$

$$= m_0c^2\left[\left\{1 - \frac{(1.8 \times 10^8)^2}{(3 \times 10^8)^2}\right\}^{-1/2} - 1\right]$$

$$= m_0c^2\,[(0.64)^{-1/2} - 1] = 0.25\, m_0c^2.$$

The final energy is

$$K_2 = m_0c^2\left[\left\{1 - \frac{(2.4 \times 10^8)^2}{(3 \times 10^8)^2}\right\}^{-1/2} - 1\right]$$

$$= m_0c^2\,[(0.36)^{-1/2} - 1] = 0.67\, m_0c^2.$$

The work required is equal to the increase in kinetic energy. Thus

$$W = K_2 - K_1$$

$$= 0.67\ m_0c^2 - 0.25\ m_0c^2 - 0.25\ m_0c^2 = 0.42\ m_0c^2$$
$$= 0.42 \times (9.1 \times 10^{-31}) \times (3 \times 10^8)^2$$
$$= 34.4 \times 10^{-15} \text{ joule}$$
$$= \frac{34.4 \times 10^{-15}}{1.6 \times 10^{-13}} = 0.215 \text{ MeV.}$$

Example 45:

Calculate the speed of an electron accelerated through a potential difference of 1.53 × 10^6 volt. Given : c = 3 × 10^8 m/s, rest mass of electron = 9.1 × 10^{-31} kg and = – 1.6 × 10^{-19} coulomb.

Solution:

According to classical mechanics, the kinetic energy of a particle of mass m_0 moving with speed v is $1/2\ m_0v^2$. When an electron accelerated through a p.d of V volt requires this speed, we have

$$\frac{1}{2}m_0v^2 = eV,$$

where e is the electron charge. From this, we get

$$v = \sqrt{\frac{2eV}{m_0}}.$$

Putting e = 1.6 × 10^{-19} C, V = 1.53 × 10^6 V and m_0 = 9.1 × 10^{-31} kg, we get

$$v = \sqrt{\frac{2 \times \left(1.6 \times 10^{-19}\right) \times \left(1.53 \times 10^6\right)}{9.1 \times 10^{-31}}} = 7.3 \times 10^8 \text{ m/s.}$$

The speed is *greater* than the speed of light (3.0 × 10^8 m/s) and hence unacceptable, because no particle can have a speed greater than the speed of light. Hence, in the given problem, we must use *relativistic* expression of kinetic energy to calculate the speed of the electron. The expression is

$$K = mc^2 - m_0c^2 = \frac{m_0c^2}{\sqrt{1-\left(\frac{v}{c}\right)^2}} - m_0c^2$$

$$= m_0c^2\left\{\frac{1}{\sqrt{1-\left(\frac{v}{c}\right)^2}} - 1\right\}.$$

When an electron acquires this energy by means of a potential V, we have

$$m_0c^2\left\{\frac{1}{\sqrt{1-\left(\frac{v}{c}\right)^2}}-1\right\}=eV$$

or
$$\frac{1}{\sqrt{1-\left(\frac{v}{c}\right)^2}}=1+\frac{eV}{m_0c^2}$$

or
$$\frac{1}{\sqrt{1-\left(\frac{v}{c}\right)^2}}=1+\frac{\left(1.6\times10^{-19}\right)\times\left(1.53\times10^{6}\right)}{\left(9.1\times10^{-31}\right)\left(3\times10^{8}\right)^2}=4$$

or
$$\sqrt{1-\left(\frac{v}{c}\right)^2}=\frac{1}{4}=0.25$$

Solving: $v = 0.968\ c.$

Example 46:

Describe the Michelson-Morley experiment and discuss the various interpretations given to the negative result obtained therefrom.

Solution:

Michelson-Morley Experiment : According to the wave theory of light, a light source sets up a disturbance travelling in all directions through a hypothetical medium called 'ether' which fills all space and penetrates all matter. The assumption of ether, however, created a problem. Does ether remain stationary in space when material bodies (including earth) move in it, or is it dragged along with the moving bodies?

Bradley's observation of the aberration of light from stars had indicated that *the ether must be stationary in space*. It means that if a material body, say earth, moves in space, there is a relative motion between the body and the ether. A number of experiments were performed to detect a relative motion between the earth and the ether. The most famous among them is the one performed by Michelson and Morley in 1887 using the Michelson interferometer. A beam of light from a source S falls upon a half-silvered glass plate P placed at 45° to the beam and is divided into two beams 1 and 2. The beams 1 and 2 travelling at right angles to each other, fall normally on mirrors M_1 and M_2 which reflect them back to P.

The two beams returned to P are directed towards a telescope T in which interference fringes are observed.

Let the mirrors M_1 and M_2 be at the same distance l from the plate P. Then, if the apparatus were at rest in ether, the two beams would take the same time to return to P. But, actually the earth, and hence the apparatus, is moving in space through the ether with a velocity v (say). Suppose this motion is in the direction of the initial beam of light. Then, if the initial beam strikes the plate P in the position shown, the paths of the two beams and the positions of their reflections from the mirrors will be as shown by the dotted lines. The time taken by the two beams on their journeys are not equal.

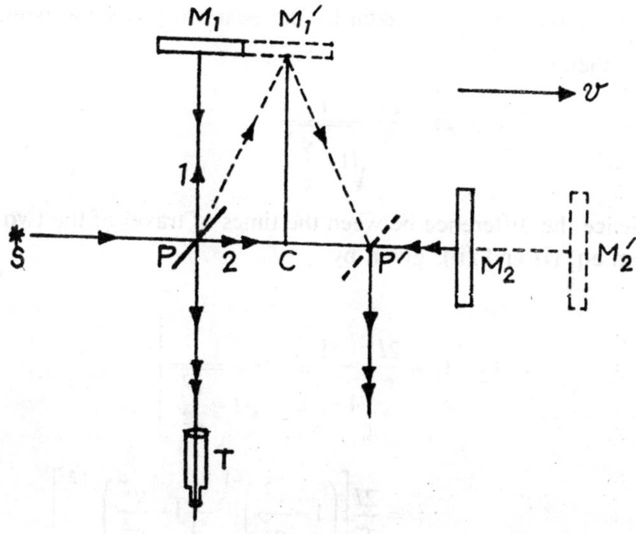

Fig. 1.12

Let c be the velocity of light through the ether. The beam 2 moving towards M_2 has a velocity (c – v) relative to the apparatus on the outgoing trip, and (c + u) on the return trip. If t_2 be the total time taken by this beam to go from P to M_2 and back, then

$$t_2 = \frac{l}{c-v} + \frac{l}{c+v} = \frac{2lc}{c^2 - v^2} = \frac{2l}{c}\left(\frac{1}{1-\frac{v^2}{c^2}}\right). \qquad ...(i)$$

The beam 1 moving *transversely* with respect to the apparatus retains its velocity c throughout. Let it take a time t' to go from P to strike M_1, travelling a distance ct'. In the same time, the mirror M_1 advances a distance vt'. Thus, in the right-angled triangle PM_1M_1', we have

$$\left(PM_1'\right)^2 = \left(PM_1\right)^2 + \left(M_1M_1'\right)^2.$$

But $PM_1 = l$, $M_1M_1' = vt'$ and $PM_1' = ct'$.

$$\therefore \qquad (ct')^2 = l^2 + (vt')^2$$

or

$$t' = \frac{l}{\left(c^2 - v^2\right)^{1/2}} = \frac{l}{c}\frac{1}{\sqrt{1-\frac{v^2}{c^2}}}.$$

If t_1 be the total time taken by the beam to travel the whole path $PM_1'P'$, then

$$t_1 = 2t' = \frac{2l}{c}\frac{1}{\sqrt{1-\frac{v^2}{c^2}}} \qquad ...(ii)$$

Hence the difference between the times of travel of the two beams is, from eq. (i) and (ii), given by

$$t_2 - t_1 = \frac{2l}{c}\left[\frac{1}{1-\frac{v^2}{c^2}} - \frac{1}{\sqrt{1-\frac{v^2}{c^2}}}\right]$$

$$= \frac{2l}{c}\left[\left(1-\frac{v^2}{c^2}\right)^{-1} - \left(1-\frac{v^2}{c^2}\right)^{-1/2}\right].$$

Using binomial expansion and dropping terms higher than the have second order, we

$$t_2 - t_1 = \frac{2l}{c}\left[\left(1+\frac{v^2}{c^2}\right) - \left(1+\frac{1}{2}\frac{v^2}{c^2}\right)\right]$$

$$= \frac{2l}{c}\left[\frac{1}{2}\frac{v^2}{c^2}\right] = \frac{lv^2}{c^3},$$

The path difference 8 between the beams corresponding to a time difference $(t_2 - t_1)$ is

$$\delta = c(t_2 - t_1) = \frac{lv^2}{c^2}.$$

If the interferometer is suddenly brought to rest (v is made zero), then the path difference δ would become zero. We know that if the path difference between two interfering waves changes by λ, there is a shift of one fringe across the cross-wires in the field of view. Thus, if ΔN is the number of fringes which shift when the interferometer is stopped, then

$$\Delta N = \frac{\delta}{\lambda} = \frac{lv^2}{c^2\lambda}.$$

In the actual experiment, the whole apparatus, which was placed on a block of stone floated on mercury, was rotated through 90°. This introduced a path difference of the same amount in the opposite direction. Hence a shift of $2lv^2/c^2\lambda$ was expected.

To have an observable shift, Michelson and Morley increased the effective value of l upto 11 meters by reflecting the light back and fourth several times. Then, using values; l = 11 meter, v = 3 × 10^4 meter/sec, c = 3 × 10^8 meter/sec and λ = 5.5 × 10^{-7} meter (for visible light), the expected shift is

$$\Delta N = \frac{2lv^2}{c^2\lambda} = \frac{2 \times 11 \times (3 \times 10^4)^2}{(3 \times 10^8)^2 \times 5.5 \times 10^{-7}} = 0.4,$$

or a shift of four-tenths a fringe.

Michelson and Morley were extremely surprised to see that there was no shift in the fringes when the interferometer was rotated through 90°. They repeated the experiment during various times of the day and various seasons of the year but no shift was observed. Trouton and Noble, in 1902, performed an electromagnetic experiment for the same purpose but with no positive result. Thus, *the motion of the earth through the ether could not be experimentally detected.*

Explanation of the Negative Result : Three separate explanations were given to the negative result of the Michelson-Morley experiment:

1. *Ether-Drag Hypothesis :* The moving earth *completely drags* the ether with it so that there is no relative motion between the two

and hence the question of shift does not arise. But this explanation was not accepted for two reasons : (i) It goes against the observed aberration of light from stars, (ii) Fizeau had experimentally shown that a moving body could drag the 'light waves' only partially. Furthermore, this partial dragging of light waves was explained by the electromagnetic theory, without introducing the ether-drag hypothesis.

2. *Fitzgerald-Lorentz Contraction Hypothesis* : Fitzgerald and Lorentz independently put an adhoc hypothesis that all material bodies moving through the ether are contracted in the direction of motion by a factor $\sqrt{1-\left(v^2/c^2\right)}$. It is easily seen that such a contraction in the interferometer arm would equalize the times t_1 and t_2 and no fringe-shift would be expected. This explanation also, being purely adhoc, could not be accepted. Further, Rayleigh worked out that such a contraction is expected to produce double refraction which was, however, never observed.
3. *Light Velocity Hypothesis* : The light from a moving source has a velocity which is the vector sum of its natural velocity and the velocity of the source. If this were true, then the light from the source had always the same constant velocity relative to the source and to the interferometer and the negative result could be explained. This explanation was also rejected because it was in conflict with the wave theory of light as well as with certain astronomical evidences concerning double stars.
4. *Einstein's Revolutionary Idea* : Einstein, in 1905, proposed a new revolutionary idea that *motion through ether is a meaningless concept; only motion relative to a frame of reference has physical significance.* The frame of reference may be a road, the earth's surface, the sun, the centre of our galaxy; but in every case we must specify it. If we were isolated in the universe, there would be no way in which we could determine whether we are in motion or not. That is why it is impossible to perform any experiment for detecting earth's motion through ether. This idea was ultimately developed in the theory of relativity.

Example 47:

Calculate the percentage contraction of a 1.0 m rod moving with a velocity of 0.8 c in a direction inclined at 60° to its own length.

Solution:

Let L_0 be the length of the rod at rest. Its components along and perpendicular to the direction of motion are $L_0 \cos 60°$ and $L_0 \sin 60°$ respectively.

The apparent length along the direction of motion

$$= L_0 60° \sqrt{1-(0.8)^2} - 0.3\,L_0.$$

and that perpendicular to the direction of motion (no change)

$$= L_0 \sin 60° = 0.866\ L_0.$$

∴ length of the moving rod is

$$L = \sqrt{(0.3\,L_0)^2 + (0.866\ L_0)^2} = 0.916\,L_0.$$

Percentage contraction

$$\frac{L_0 - L}{L_0} \times 100 = \frac{L_0 - 0.916\,L_0}{L_0} \times 100 = 8.4\%$$

Example 48:

The area of a certain square is L_0^2 as measured by an observer at rest. If another observer is moving with a velocity v relative to the first and parallel to one of the sides of the square, then calculate the area of the square as measured by the second observer.

Solution:

According to a consequence of Lorentz transformation equations, the length of a body moving with a velocity v relative to the observer is contracted by a factor $\sqrt{1-\left(v^2/c^2\right)}$ in the .direction of its motion, whereas its dimensions perpendicular to the direction of motion are unaffected.

Thus, the square, the rest length of each side of which is L_0, has its length along the direction of motion as

$$L_0\left(1-\frac{v^2}{c^2}\right)^{1/2},$$

while the length in the perpendicular direction is still L_0. Hence the area when in motion is

$$L_0^2\left(1-\frac{v^2}{c^2}\right)^{1/2}.$$

EXERCISES

24. Write a brief account of the stationary ether hypothesis, explaining clearly why the medium ether was invented in the first place and discarded later.
25. What change in mass is associated, in a chemical reaction with the:
    (a) absorption
    (b) release, of 1eV of energy?
26. An electron and a positron which have negligible velocities combine to produce two-photon annihilation radiation:
    (a) What is the energy of each photon?
    (b) What will be the relative direction of motion of these photons? Explain.
27. The fissioning of an atom of uranium–235 releases 200 MeV of energy. What percent is this fission energy of the total which would have been available if all the mass of the uranium atom had appeared as energy?
28. How fast would a rocket ship have to go relative to an observer for its length to be contracted to 99% of its length when at rest?
29. The length of the side of a square, as measured by an observer in a stationary frame of reference S, is *l*. What will be its apparent area, as observed by him in a reference frame S' moving relative to S with velocity v along one of the sides of the square?
31. The hypothetical speed of the earth through the ether is its orbital speed of 3×10^4 m/sec. If the light takes 3×10^{-7} sec to travel through the Michalson-Morley apparatus in the direction parallel to this motion, how long will it take to travel through it in the direction perpendicular to this motion?
32. Explain the basic postulates of Einstein's special theory of relativity. Derive the Lorentz space time transformation formulae.
33. A frame of reference S' is moving relative to a frame S with velocity 0.6 ci. A vector in S' is represented by 5i + 8j + 4k. How may it be represented in frame 5?

34. A rod of true length 100 cm is moving with velocity 0.6c in a direction making an angle of 30° with its length. What is the contraction produced in its length and along what direction does it appear to move?

35. What do you understand by lime-dilation? What is proper interval of time? Briefly discuss one experiment in support of time dilation in special relativity.

5. What is the meaning of mass-energy equivalence? Obtain Einstein's mass-energy relation. Show that 1 amu = 931 MeV.

6. Show that the expression $E_k = \frac{1}{2}mv^2$ does not give the relativistic value of the kinetic energy of a body even if m represents its relativistic mass.

7. State the basic postulates of the special theory of relativity and hence obtain Lorentz transformation. Discuss length contraction and time dilation.

8. How fast should a rocket ship move relative to an observer in order that one year on it may correspond to two years on the earth?

9. Write a note on Einstein's mass-energy relation. What is the principle of mass and energy equivalence? Explain it by giving examples.

10. Calculate the wavelength of the radiation emitted by the annihilation of an electron with a positron, each of rest mass 9.1×10^{-28} gm.

11. A proton has a velocity equal to 0.999c in the laboratory. Find the energy and momentum as observed in a frame travelling in the same direction with velocity s0.990 c with respect to the laboratory.

13. Show that the relativistic equations to transform space and time to system S from observations on system S' are $x = y(x' + vt')$ and $t = y(t' + vx'/c^2)$ respectively, where S' is moving relative to S with velocity v along the axis of x.

19. State and deduce the mathematical expression for the law of addition of relativistic velocities. Show that in no case can the resultant velocity of a material particle be greater than c.

    Also show that the Lorentz velocity transformation equations reduce to Galilian ones for values of $v << c$.

20. Two electrons move towards each then the speed of each being 0.9c in a Galilian frame of reference. What is their speed relative to each other?
21. In the laboratory, electrons from two accelerators are projected with the same speed of 2×10^8m/sec but in opposite directions. What is the relative velocity of the two sets of electrons?
22. Describe the Michelson-Morley experiment and explain the physical significance of the negative results.
23. Calculate the expected fringe shift in the Michelson-Morley experiment if the effective length of each path be 6 metres, velocity of the earth, 3×10^4m/sec and the wavelength of the monochromatic light used, 5000 Angstrom units.

2

Interference of Light

INTRODUCTION

The result is then called interference of waves. Two equal corpuscles (of Newton) reaching a point in space would always mean 2-fold intensity, but two equal waves (of Huygens) reaching a point in space can mean anything from 4-fold to zero intensity, depending on the phase relation between the two waves. That is the crux of interference. Huygens did not know anything about the nature of the light wave; whether it is a transverse wave or a longitudinal wave; he had no knowledge about the speed of light or its wavelength. The wave theory of light received the first experimental evidence in 1801 from the interference experiments conducted by Thomas Young. Using the principle of superposition and Huygens' wave concept. The Young's experiment, Fresnel's biprism, Lloyd's mirror, etc., the coherent sources are obtained by dividing the wave-front in width.

Another way of producing coherent beams with light is to use multiple reflection between two faces of a transparent material. In this case widths of the successively transmitted as well as the reflected wave-fronts are about the same as that of the incident one; but the amplitudes are different. A number of coherent beams—not just two—are produced on both sides of a film by a single incident beam, and these beams show interference. Thus, here we have coherent beams produced by division of amplitude. Note that the basis is that at any boundary between two different media there is always part-reflection and part-transmission.

PLANE PARALLEL FILM

A transparent thin film of uniform thickness bounded by two parallel surfaces is known as a *plane parallel thin film.*

When light is incident on a parallel thin film, a small portion of it gets reflected from the top surface and a major portion is transmitted into the film. Again, a small part of the transmitted component is reflected back into the film by the bottom surface and the rest of it is transmitted from the lower surface of the film. Thin films transmit incident light strongly and reflect only weakly. After two reflections, the intensities of reflected rays drop to a negligible strength. Therefore, we consider the first two reflected rays only (Fig. 2.1). These two rays are derived from the same incident ray but appear to come from two sources located below the film. The sources are virtual coherent sources. The reflected waves 1 and 2 travel along parallel paths and interfere at infinity. This is a case of *two-beam* interference.

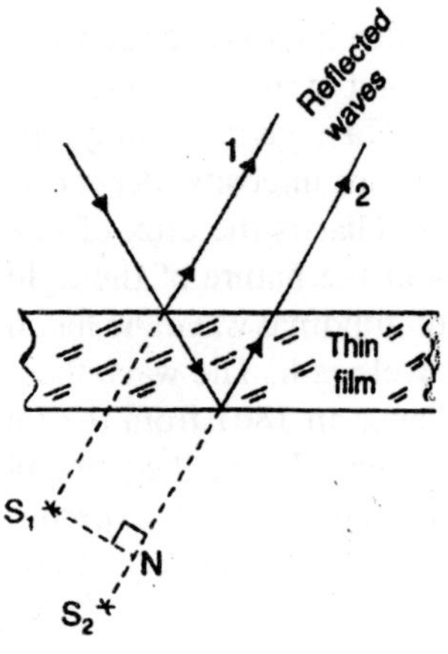

Fig. 2.1

The condition for maxima and minima can be deduced once we have calculated the optical path difference between the two rays at the point of their meeting.

(i) Geometrical Path Difference

Let DH be normal to BC. From points H and D onwards, the rays HC and DE travel equal path. The ray BH travels in air while the ray BD travels in the film of refractive index μ along the path BP and FD. The geometric path difference between the two rays is

$$BP + PD - BH.$$

(ii) Optical Path Difference

Optical path difference $\Delta_a = \mu\ L$

$\therefore\ \Delta_a = \mu\ (BF + FD) - 1\ (BH)$ ...(1)

$\therefore$ In the ΔBFD, $\angle BFG = \angle GFD = \angle r$

$$BP = FD$$

$$BF = \frac{FG}{\cos r} = \frac{t}{\cos r}$$

$$\therefore \quad BF + FD = \frac{2t}{\cos r} \quad ...(2)$$

Also,

$$BG = GD$$

$$\therefore \quad BD = 2BG$$

$$BG = FG \tan r = t \tan r$$

$$BD = 2\ t \tan r$$

In the Δ^{le} BHD, $\angle HBD = (90 - i)$

$$\angle BHD = 90°$$

$$\therefore \quad \angle BDH = i$$

$$BH = BD \sin i = 2\ t \tan r \sin i \quad ...(3)$$

From Snell's law,

$$\sin i = \mu \sin r$$

$$\therefore \quad BH = 2t \tan r\ (\mu \sin r) = \frac{2\mu t \sin^2 r}{\cos r} \quad ...(4)$$

Using the equations (2) and (4) into eqn. (1), we get

$$\Delta_a = \mu \left[\frac{2t}{\cos r}\right] - \left[\frac{2\mu t \sin^{2 r}}{\cos r}\right]$$

$$= \frac{2\mu t}{\cos r}\left[1 - \sin^2 r\right]$$

$$= \frac{2\mu t}{\cos r} \cos^2 r$$

$$\therefore \quad \Delta_a = 2\mu\ t \cos r \quad ...(5)$$

(iii) Correction on Account of Phase Change at Reflection

When a ray is reflected at the boundary of a rarer to denser medium, a path change of $\lambda/2$ occurs for the ray BC. There is no path difference due to transmission at D. Including the change in path difference due to reflection, the true path difference

$$\Delta_t = 2\,\mu\,t\cos r - \frac{\lambda}{2} \qquad ...(6)$$

Conditions for Maxima (Brightness) and Minima (Darkness)

Maxima occur when the optical path difference $\Delta = m\lambda$. If the difference in the optical path between the two rays is equal to an *integral number of full waves*, then the rays meet each other in phase. The crests of one wave falls on the crests of the others and the waves *interfere constructively*.

Thus, when

$$2\mu\,t\cos r - \frac{\lambda}{2} = m\lambda \qquad ...(7)$$

the reflected rays undergo constructive interference to produce brightness or maxima at the point of their meeting.

$$2\mu\,t\cos r = m\lambda + \lambda/2$$

or $\quad 2\mu\,t\cos r = (2m + 1)\lambda/1$ *(Condition for Brightness)* ...(8)

Minima occur when the optical path difference is $\Delta = (2m + 1)\lambda/2$. If the difference in the optical path between the two rays is equal to an *odd integral number of half-waves*, then the rays meet each other in opposite phase. The crests of one wave falls on the troughs of the others and the waves *interfere destructively*. Thus, when

$$2\mu\,t\cos r - \lambda/2 = (2m + 1)\,\lambda/2 \qquad ...(9)$$

the reflected rays undergo destructive interference to produce darkness. Eqn. (9) may be rewritten as

$$2\mu\,t\cos r = (m + 1)\,\lambda$$

The phase relationship of the interfering waves does not change if one full wave is added to or subtracted from any of the interfering waves. Therefore $(m + 1)\,\lambda$, can as well be replaced by mk for simplicity in expression. Thus,

$2\mu\,t\cos r = m\lambda$ *(Condition for Darkness)* ...(10)

Some Important Points

(a) It is seen that the conditions of interference depend on four parameters, namely μ, t, λ and r. In the case of constant thickness (parallel) film, (μt) is constant. When a parallel, beam of light is incident on such a film, r also remains constant. Then the interference conditions solely depend on the wavelength λ.

(b) When monochromatic light falls on a parallel beam, the whole film will appear *uniformly* dark or bright. If the condition of constructive interference is satisfied, the film will show intense colour corresponding to the incident light.

(c) If a parallel beam of white light falls on a parallel film, those wavelengths for which the path difference is $m\lambda$, will be absent from the reflected light. The other colours will be reflected. Therefore, the film will appear uniformly coloured with one colour being absent.

ELECTROMAGNETIC WAVES

The light wave is a harmonic electromagnetic wave consisting of periodically varying electric and magnetic fields oscillating at right angles to each other and also to the direction of propagation of the wave (Fig. 2.2a). The electric field in the wave is defined by the electric field strength vector E and the magnetic field by the vector of magnetic induction B. Vectors E and B are of equal importance to the wave. However, a light wave is often represented by the E wave (Fig. 2.2b), since many of the effects of light such as photoelectric effect, photochemical and physiological actions are found to be mostly due to the electric vector E. The vector E is often referred to as the *light vector* or *optical vector*. The electric field is known as *optical field*, radiation field, wave field or light field. The magnetic field is implied to be oscillating in a plane normal to the plane of the electric field oscillations and is not shown specifically in the diagrams.

The light wave depicted in Fig. 2.2(b) is mathematically represented by the expression

$$E = E_0 \sin (k z - \omega t). \qquad ...(1)$$

In fact, the wave shown in Fig. 2.2 (b) and represented by eqn. (1) is an ideal electromagnetic wave. *No real light source emits such perfect waves.*

1. We note the following features about such a mathematical wave.

   (i) The wave has a single definite frequency ν ($= \omega/2\pi$). In optics, waves having a single frequency and wavelength are called *monochromatic* waves.

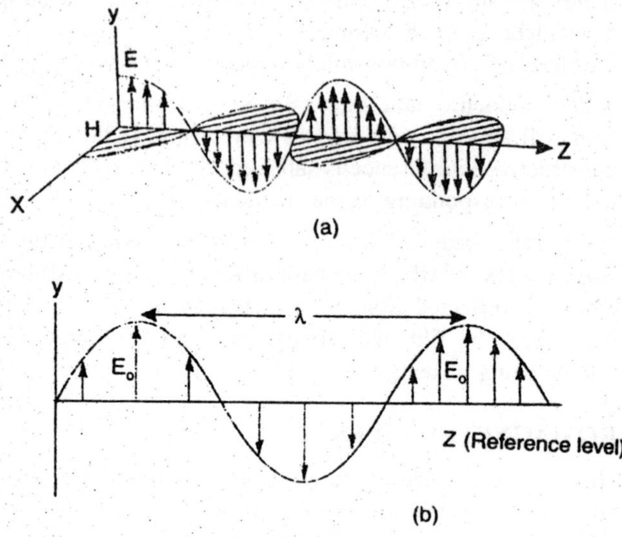

Fig. 2.2

(ii) It is a *harmonic* wave and is of *infinite extension* consisting of a continuous *train of waves*. At any instant the wave extends from $z = -\infty$ to $z = +\infty$, and at any point whose position corresponds to a particular value of z the wave continues from $t = -\infty$ to $t = +\infty$.

(iii) The amplitude of the wave E_0 stays constant as the wave propagates through air. Hence it is a plane wave and its wavefront is normal to the z-axis.

(iv) The electric vector E of the wave oscillates always parallel to a fixed direction in space, *i.e.*, y-direction. In other words, the E-vibrations are confined to yz – plane. Therefore the wave is *plane polarized* (or *linearly polarized*).

The real light waves emitted by common light sources are far from ideal. They are wave trains of limited length, have a spread of frequencies, and are not plane polarized. Interference effects are easily observed when sinusoidal waves with a single frequency, v, and wavelength, λ, overlap. While it is fairly easy to produce sound waves and r.f. waves of single frequency, common sources of light *do not emit* monochromatic light. The common light sources emit a continuous distribution of wavelengths. The sodium lamp used in optics laboratories is a fairly monochromatic source of light, which emits light in the yellow region at the wavelength 5893 Å. The most nearly monochromatic source is the helium-neon laser that emits red light at 6328 Å.

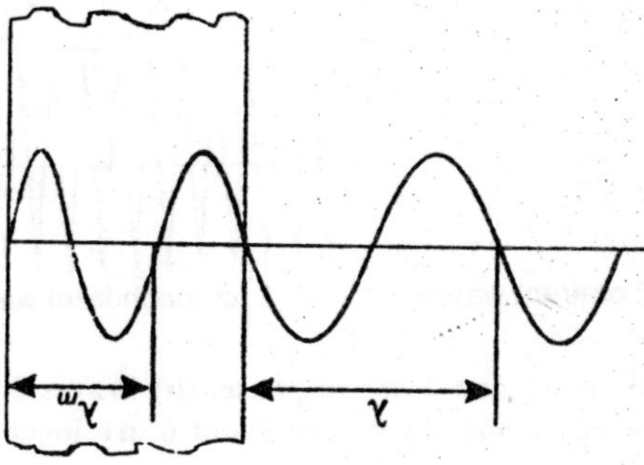

Fig. 2.3

2. A light wave travels *slower* in an optical medium than in air or a vacuum. It travels with a velocity υ, which is less than 'c'. The wavelength of light wave *decreases in the medium*, as shown in Fig. 2.3 while its frequency remains constant.

3. *Phase difference and coherence :* Suppose two waves are passing through a point in space. If the frequencies of the two waves are different, the phase difference between the vibrations changes with time. The waves will drift out of phase because the crests of the higher frequency wave will arrive ahead of the crests of the lower frequency wave. Also, if one (or both) of the waves undergoes changes in frequency irregularly, the phase difference changes irregularly. Under these conditions the two waves are said to be *incoherent.*

   The light emitted by most of the light sources is incoherent as the frequency of light changes abruptly and irregularly, though we can think of an average frequency associated with the light wave. On the other hand, if we consider two waves of same frequency, they may differ in the amplitudes but they maintain a predictable phase relationship (Fig. 2.4a). The difference in their phases may have any value from zero radians to a maximum of 2π radians; but the phase difference remains constant. Thus, two or more waves of the same frequency can maintain the same phase or constant phase difference over a distance and time. Such waves are said to be *coherent waves.*

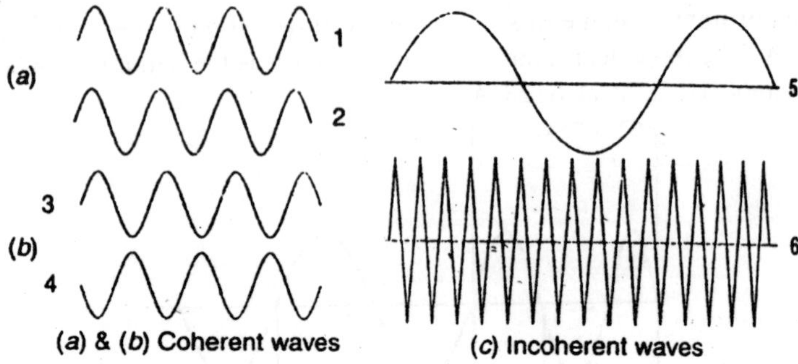

Fig. 2.4 : (a) Waves 1 and 2 are in-phase, (b) Waves 3 and 4 are in opposite phase and (c) Waves 5 and 6 are incoherent.

When coherent waves rise or fall together, reaching the crest (or trough) at the same time, they are said to be *in phase*, as shown in Fig. 2.4 (a). The phase difference will be 0 or 2n radians, which remains constant as the waves propagate in space. The path difference between the waves will then be zero or an integral multiple of a wavelength, λ. Such waves move through space with a *crest-to-crest correspondence*. When one wave reaches its crest while the other falls to its trough, then the phase difference between the waves is π and the waves are said to be in *opposite phase* and the phase difference remains 180° (Fig. 2.4b). They are inverted with respect to each other everywhere. Then the path difference between the waves will be λ/2 or an odd integral multiple of π/2.

The waves may have also any constant *phase difference* other than zero or π radians.

If two (or more) waves maintain a constant phase difference over a long distance and time, then they are said to be coherent.

Two waves of different frequency (Fig. 2.4c) can never maintain a constant phase difference, because their phase difference goes on fluctuating and changes arbitrarily. They are said to be *incoherent*, as the phase difference between the waves fluctuates with time.

Thus, *coherence* means the coordinated motion of several waves in a medium maintaining a fixed and predictable phase relationship over a length of time.

The waves shown in Fig. 2.4 (a) and (b) are coherent waves. Sources, which produce coherent waves, are called *coherent sources*.

In the study of interference, we focus our attention mainly on the two specific types of disposition of the waves, namely in phase and opposite phase dispositions, at the point of observation.

4. *Optical path and Phase change :* Optical path length indicates the number of light waves that fit into that path. Thus, $\Delta = N\lambda$, where Δ is the optical path length and N is an integer or a mixed fraction. Optical path is related to the geometric path and the relation may be found as follows. The distance traversed by light in a medium of refractive index μ in time t is given by

$$L = \upsilon t \qquad ...(2)$$

where υ is the velocity of light in the medium. The distance travelled by light in a vacuum in the same time t, is $\Delta = c\,t = c\,\dfrac{L}{\upsilon} = \mu L$

The distance L is called the *geometric path* length (GPL). A is the equivalent distance in a vacuum and is called *optical path length* (OPL). Thus,

$$\text{O.P.L.} = \mu \times \text{G.P.L.}$$

or

$$\Delta = \mu L \qquad ...(3)$$

The above result means that *more number of waveforms is accommodated along the optical path than in the corresponding geometrical path.* Therefore, in the study of interference we always must calculate the optical paths travelled by light rays.

(a) *Effect of Optical path :* Optical path determines the *phase* of a light wave arriving at a point. We know that if a wave covers in air a distance of one wavelength, *i.e.* 1λ its phase changes by 2π radians. Therefore, we compute that if a wave travels a distance L in air, its phase change is given by

$$\delta = \frac{2\pi L}{\lambda} \qquad ...(4)$$

When the wave travels the distance L in a medium, then

$$\delta = \frac{2\pi\Delta}{\lambda} = \frac{2\pi\mu L}{\lambda}$$

where Δ is the optical path or optical path difference.

Comparing eqns. (4) and (5), we find that a light path of geometric length L in a medium of refractive index u produces the same phase change as a light path of length μL in a vacuum.

(b) *Effect of reflection* : The process of reflection also affects the phase of a light wave. When light is incident on a surface, part of the light gets reflected while a major portion may be *transmitted or absorbed.* The quantity characterizing the reflectivity of a surface is called the reflection coefficient, ρ.ρ depends on the nature of surface, the angle of incidence of light and many other factors. Augustin Jean Fresnel, about 150 years ago, derived a set of expressions which allow us to calculate the amount of light reflected and transmitted at an interface. For normal incidence it was shown that

$$\rho = \frac{\mu_1 - \mu_2}{\mu_1 + \mu_2}$$

It may be seen that ρ is positive when $\mu_2 < \mu_1$. It implies that the oscillations in the incident and reflected waves occur in the same phase. On the other hand, when $\mu_2 > \mu_1$, ρ is *negative* signifying that the *oscillations in the incident and reflected waves are in opposite phase.* Hence, we draw the following conclusions:

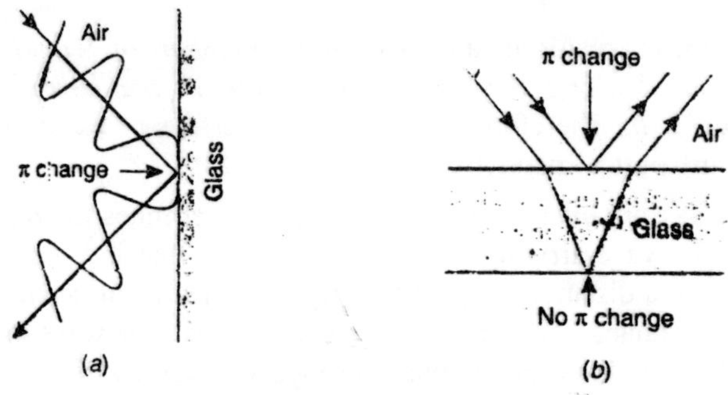

Fig. 2.5 : Phase change due to reflection.

(i) A light wave travelling from a rarer medium (μ_1) to a denser medium (μ_2) undergoes a phase change of π radians when it gets reflected at the boundary of denser medium, as shown in Fig. 2.5 (a). The wave loses a half-wave on reflection at the boundary of rarer-to-denser medium.

(ii) A light wave travelling from a denser medium (μ_2) to a rarer medium (μ_1) does not undergo a change in phase on reflection at the boundary of denser-to-rarer medium (Fig. 2.5b). Therefore, the change in path is zero.

SUPERPOSITION OF WAVES

Frequently it is necessary to find the resultant disturbance at a point when a number of disturbances arrive simultaneously. According to the *principle of superposition—*

when two or more waves overlap, the resultant displacement at any point and at any instant may be found by adding the instantaneous displacements that would be produced at the point by the individual waves if each were present alone.

It means that the resultant is simply the sum of the disturbances. The principle of superposition applies to electromagnetic waves also and is the most important principle in wave optics. In case of electromagnetic waves, the term *displacement* refers to the amplitude of the electric field vector.

Interference is an important consequence of superposition of *coherent waves.*

INTERFERENCE DUE TO TRANSMITTED LIGHT

Consider a thin transparent film of thickness t and refractive index μ A ray SA after refraction goes along AB. At B it is partly reflected along BC and partly refracted along BR. The ray BC, after reflection at C, finally emerges along DQ. Here at B and C reflection takes place at the rarer medium. Therefore, no phase change occurs. Draw BM normal to CD and DN normal to BR. The optical path difference between DQ and BR is given by

$$\Delta = \mu(BC + CD) - BN$$

Also, $\mu = \dfrac{\sin i}{\sin r} = \dfrac{BN}{MD}$ or $BN + \mu.MD$

In Fig. 15.7, $\angle BPC = r$ and $CP = BC = CD$

$\therefore \quad BC + CD = PD$

$\therefore \Delta = \mu(PD) - \mu(MD) = \mu(PD - MD) = \mu.PM$

In ΔBPM, $\cos r = \dfrac{PM}{BP}$ or $PM = BP.\cos r$

But $\quad BP = 2t$

$\therefore \quad PM = 2t \cos r$

$\therefore \quad \Delta = \mu.PM = 2\mu t \cos r \qquad ...(1)$

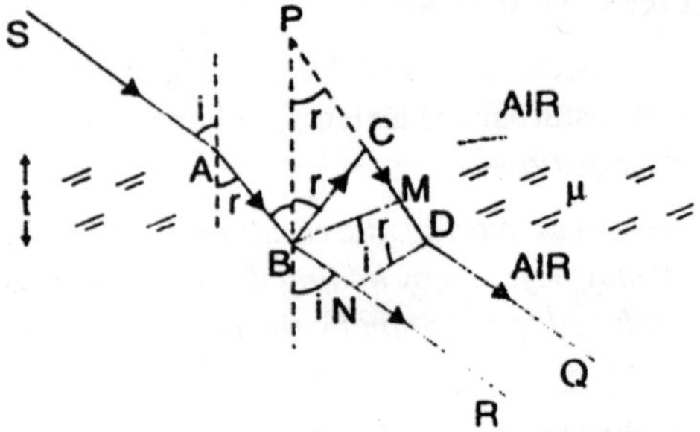

Fig. 2.6

Bright Fringes

When the optical path difference $\Delta = \Delta\ \lambda$, bright fringe occurs.

$\therefore \quad 2\Delta\ t \cos r = m\lambda$ ...(2)

where m = 0, 1, 2, 3,....

Dark Fringes

When the optical path difference $\Delta = (2m + 1)\ \lambda/2$, dark fringe occurs.

$$\therefore \quad 2\mu\ t \cos r = \frac{(2m+1)}{2}\lambda \qquad ...(3)$$

where m = 0, 1, 2, 3,....

In case of transmitted light, the fringes are less distinct because the difference in amplitudes of BR and DQ is very large. However, when the angle of incidence is nearly 45° the fringes are more distinct.

HAIDINGER FRINGES

In thin films interference fringes are produced due to the path difference $2\mu\ t \cos r$ between the overlapping rays. For a given film the path difference may arise due to

(i) the angle of refraction r inside the film or

(ii) the change in thickness.

We can express the change in path difference by differentiating the expression $2\ \mu\ t \cos r$.

Change in path difference,

$$\delta(\Delta) = 2\mu\ t.\delta\ (\cos r) + 2\mu \cos r\ (\delta t) \qquad ...(1)$$

When the film is of *uniform* (constant) thickness, the change in path difference is only due to the change in r. If the thickness of the film is large, the path difference will change appreciably even when r changes in a small way. Fringes are produced in this case due to the superposition of rays, which are equally inclined to the normal. These fringes are called *fringes of equal inclination.* The fringes of equal inclination are known as *Haidinger fringes.* In this case all the pairs of interfering rays of equal inclination pass through the plate as a parallel beam and hence meet at infinity. The other pairs of different inclination meet at different points at infinity. Therefore, they can be located with a telescope focussed to infinity. The fringes are therefore said to be *localized at infinity.*

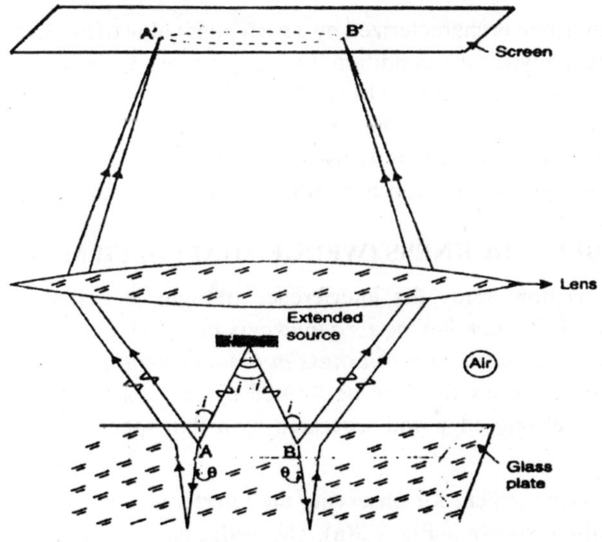

Fig. 2.7

To produce Haidinger fringes, the source must be an extended source, the film thickness must be appreciably large and the observing instrument is to be focussed for parallel rays.

Fig. 2.7 shows the formation of Haidinger fringes. Let us consider that a thin plate is illuminated by an extended monochromatic light source. A lens is arranged parallel to the plate and a screen is kept in the focal plane of the lens. Light from the extended source is incident on the plate in diverse directions. The waves propagating parallel to the plane of the page and falling on the plate at an angle ; at points A and

B get reflected from the top and bottom surfaces of the plate. The reflected pairs of waves will meet at points A' and B' respectively on the screen due to the focussing action of the lens. Depending on their path difference, the reflected waves produce either brightness or darkness on the screen. In fact the waves incident at the top surface of the plate at an angle i travel along the generators of a cone as shown in Fig. 15.8. Each pair of parallel reflected waves interfere at diametrically opposite points. Thus, a circular fringe is produced. Similarly, the waves incident at a different angle will produce a collection of identical points arranged along a circle of another radius. As a result, a system of alternating bright and dark circular fringes with a common centre will be observed on the screen.

Each fringe is characterized by a particular value of m. Bright fringes are produced when the condition $2 \mu t \cos r = m\lambda$ is satisfied; and dark fringes are produced where the condition $2 \mu t \cos r = (2m + 1) \mu/1$ is satisfied. The parallel pairs of reflected rays meet only at infinity; therefore a lens is used to focus them. Accordingly, these fringes of equal inclination are said to be localized at infinity.

VARIABLE THICKNESS (WEDGE-SHAPED) FILM

Let us now study the interference of light in a film of varying thickness. *A thin film having zero thickness at one end and progressively increasing to a particular thickness at the other end is called a wedge.* A thin wedge of air film can be formed by two glass slides resting on each other at one edge and separated by a thin spaces at the opposite edge.

The arrangement for observing the interference pattern in a wedge shaped film is shown in Fig. 2.8(a). The wedge angle is usually very small and of the order of a fraction of a degree. When a parallel beam of *monochromatic* light illuminates the wedge from above, the ray reflected from its two bounding surfaces will not be parallel. They appear to diverge from a point near the film. The path difference between the rays reflected from the upper and lower surfaces of the air film varies along its length due to variation in film thickness. Therefore, alternate bright and dark fringes are observed on its top surface (Fig. 2.8b). The fringes are localized at the top surface of the film.

When the light is incident on the wedge from above, it gets partly reflected from the glass-to-air boundary at the top of the air film. Part of the light is transmitted through the air film and gets reflected partly at the air-to-glass boundary, as shown in Fig. 2.9. The two rays BC and DE, thus reflected from the top and bottom of the air film, are coherent as they are derived from the same ray AB through division of *amplitude*. The rays are close enough if the thickness of the film is of the order of a wavelength of light. For small film thickness the rays interfere producing darkness or brightness depending on the phase difference. The thickness of the glass plates is large compared with the wavelength of the incident light. Hence, the observed interference effects are entirely due to the wedge-slipped air film.

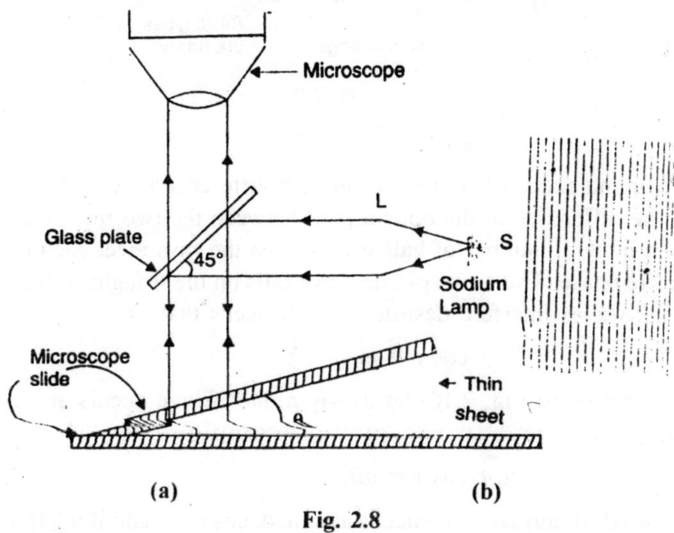

Fig. 2.8

The optical difference between the two rays BC and DE is given by

$$\Delta = 2\mu \, t \cos r - \lambda/2$$

where $\lambda/2$ takes account the gain of half-wave due to the abrupt jump of π radians in the phase of the wave reflected from the bottom boundary of air-to-glass.

Maxima occur when the optical path difference $\Delta = m\lambda$. If the difference in the optical path between the two rays is equal to an *integral number of full waves*, then the rays meet each other in phase. The crests of one wave falls on the crests of the others and the waves *interfere constructively*. This needs that

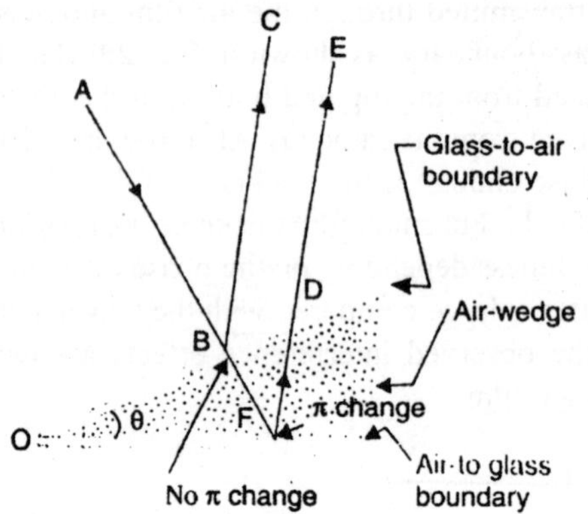

Fig. 2.9

$$\Delta = 2\mu\ t \cos r - \lambda/2.$$

Minima occur when the optical path difference is $\Delta = (2\ m + 1)\ \lambda/2$. If the difference in the optical path between the two rays is equal to an odd integral number of half-waves, then the rays meet each other in opposite phase. The crests of one wave falls on the troughs of the others and the waves interfere destructively. It needs that

$$2\mu\ t \cos r = m\lambda$$

Referring to Fig. 2.10, let us say a dark fringe occurs at A where the relation

$$2\mu\ t \cos r = m\lambda$$

is satisfied. If normal incidence is assumed, cos r = 1 and if the thickness of air film at A is denoted by t_1 then at A

$$2\mu\ t_1 = m\lambda \qquad ...(2)$$

The next dark fringe will occur, say, at C

where the thickness CL = t_2 Then at C

$$2\mu\ t_2 = (m + 1)\lambda \qquad ...(3)$$

Subtracting eqn. (2) from eqn. (3), we get

$$2\mu.(t_2 - t_1) = \lambda \qquad ...(4)$$

But $(t_2 - t_1)$ = BC

$\therefore$ $2\mu(BC) = \lambda$

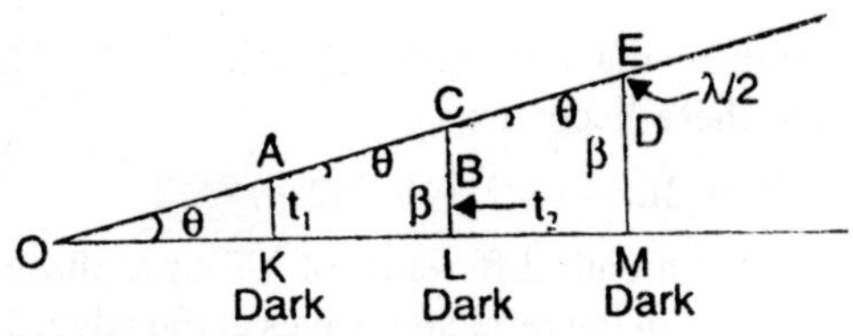

Fig. 2.10

or $$BC = \frac{\lambda}{2\mu} \qquad ...(5)$$

From the $\Delta^{le}ABC$, $\angle CAB = \theta$ and $BC = AB \tan\theta$

$$\therefore \qquad (AB)\tan\theta = \frac{\lambda}{2\mu} \qquad ...(6)$$

AB is the distance between successive dark fringes and it also equals the separation of the

successive bright fringes. It is, therefore, called the *fringe width, β*. That is AB = β. We may write eqn. (6) as

$$\beta = \frac{\lambda}{2\mu\tan\theta} \qquad ...(7)$$

For small values of θ, $\tan\theta \approx \theta$.

$$\therefore \qquad \beta = \frac{\lambda}{2\mu\theta} \qquad ...(8)$$

As the quantities on the right side of the above equation are all constant, β is constant for a given wedge angle. According to eqn. (8), an increase in the angle θ makes the fringes move closer. At an angle $\theta \approx 1°$, the interference pattern vanishes. On the other hand, if θ is gradually decreased, the fringe separation increases, and ultimately the fringes disappear as the faces of the film become parallel.

The interference pattern has the following salient features.

(i) Fringe at the apex is dark.

(ii) Fringes are straight and parallel.

(iii) Fringes are equidistant.

(iv) Fringes are localized.

(v) Fringes are of equal thickness.

(i) *Fringe at the Apex is Dark :* At the apex, the two glass slides are in contact with each other. Therefore, the thickness of the air film at the contact edge is negligible ($t \cong 0$). The optical path difference there becomes

$$\Delta = 2\mu t - \lambda/2 = 0 - \lambda/2 = -\Delta/2 \quad ...(9)$$

It implies that a path difference of $\lambda/2$ or a phase difference of n occurs between the reflected waves at the edge. The two waves interference destructively. Therefore, the fringe at the apex is always dark (Fig. 2.11).

(ii) *Straight and Parallel Fringes :* Each fringe in the pattern is produced by the interference of rays reflected from sections of the wedge having the same thickness. The locus of points having the same thickness lie along lines parallel to the contact edge. Therefore, the fringes are straight. Since the fringes are equidistant, they will be parallel (Fig. 2.11).

(iii) *Equidistant Fringes :* The fringe width P is given by

$$\beta \approx \lambda/2\theta \quad ...(10)$$

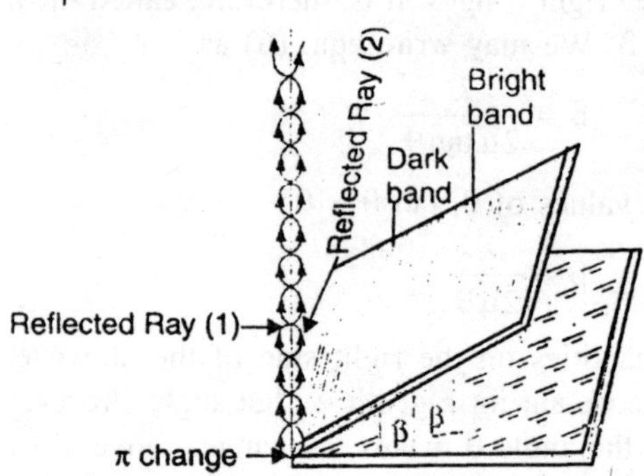

Fig. 2.11

where λ is the wavelength of the incident light and θ is the angle of the wedge. As the quantities λ and θ are constants, β is constant for a given wedge angle. Therefore, the fringes are equidistant (Fig. 2.11).

(iv) *Localized fringes :* The fringes form very close to the top surface of the wedge and can be seen with a microscope.

(v) *Fringes of equal thickness :* In thin films of thickness of the order of a few λ, the rays from various parts of the film have almost the same inclination and hence the path difference between the overlapping waves changes mainly due to change of thickness.

The fringes produced in such cases are mainly due to the variation in thickness of the film. Each fringe will be the locus of points of the same thickness. Such fringes are called fringes of equal thickness.

Determination of the Wedge Angle

The wedge angle θ can be experimentally determined with the help of a travelling microscope. Using the microscope the positions of dark fringes at two distant points Q and R are noted (Fig. 2.12). Let the distance OQ be x_1 and OR be x_2. Let the thickness of the wedge be t_1 at Q and t_2 at R.

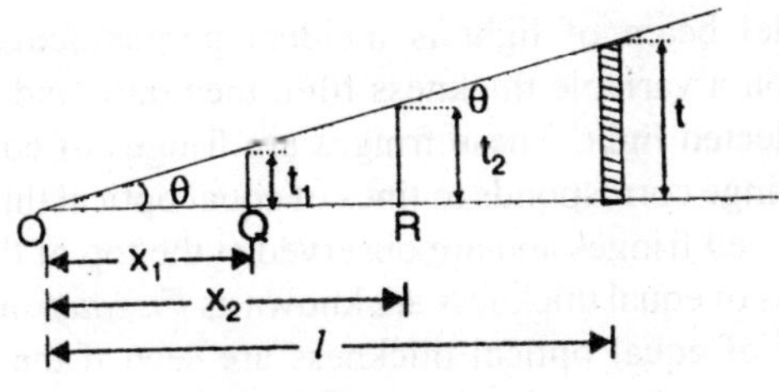

Fig. 2.12

The dark fringe at Q is given by

$$2\mu t_1 = m\lambda \qquad ...(11)$$

But as θ is very small, we can write

$$t_1 = x_1 \tan\theta = x_1\theta$$

$$\therefore \quad 2\mu x_1 \theta = m\lambda \qquad ...(12)$$

We can write similarly for the dark fringe at R as

$$2\mu x_2 \theta = (m + N)\lambda \qquad (13)$$

where N is the number of dark fringes lying between the positions Q and R. Subtracting eqn. (12) from eqn. (13), we get

$$2\mu(x_2 - x_1)\theta = N\lambda$$

$$\therefore \quad \theta = \frac{N\lambda}{2\mu(x_2 - x_1)} \qquad ...(14)$$

In case of air $\mu = 1$ and the above relation reduces to

$$\theta = \frac{N\lambda}{2(x_2 - x_1)}$$

Determination of the Thickness of the Spacer

The thickness of the spacer used to form the wedge shaped air film between the glass slides can be determined from the above measurements. If 't' is the thickness of the spacer (foil or wire) used, we can write from Fig. 2.12 that

$$t = l \tan \theta \cong l\,\theta \qquad ...(15)$$

where I is the length of the air wedge. Using the eqn. (15), we obtain

$$\therefore \quad t = \frac{lN\lambda}{2(x_2 - x_1)} \qquad ...(16)$$

Fizeau Fringes

If a parallel beam of light is incident perpendicularly or nearly perpendicular on a variable thickness film, then dark and bright fringes are seen in reflected light. These fringes are fringes of equal thickness, because each fringe corresponds to lines of equal optical thickness. These fringes or localized fringes and are observed at the top of the film. These localized fringes of equal thickness are known as *Fizeau fringes*. Contours following lines of equal optical thickness are seen if the area is large. The fringes may be obtained in case of thick films also if the source is small.

Colours in Thin Films

The colours exhibited in reflection by thin films of oil, mica, soap bubbles and coatings of oxides on heated metals etc. are due to interference of light from an extended source such as sky. Thomas Young explained the origin of colours in thin films. It may be understood as follows. The films are usually observed by reflected light. The eye looking at the thin film receives light waves reflected from the top and bottom surfaces of the film. The reflected rays are very close to each other and are in a position to interfere. The optical path difference between the interfering rays is $\Delta = 2\mu t \cos r - \lambda/2$. It is seen that the path difference depends upon the thickness t of the film the wavelength λ, and the angle r, which is related to the angle of incidence of light on the film. White light consists of a range of wavelengths and for specific values of t and r, waves of only certain wavelengths (colours) constructively interfere. Therefore, only those colours are present in the reflected light. The other wavelengths interfere destructively and hence are absent from the reflected light. Hence, the film at a particular point appears coloured. As the thickness and the angle of incidence vary from point to point, different colours are intensified at different places. The colours seen are not isolated colours, as at each place there is a mixture of colours. The composition of colours is different at different places and contours of impressive hues are observed over the entire surface of the film.

INTERFERENCE

If two or more light waves of the same frequency overlap at a point, the resultant effect depends on the *phases* of the waves as well as their *amplitudes*. The resultant wave at any point at any instant of time is governed by the *principle of superposition*. The combined effect at each point of the region of superposition is obtained by adding algebraically the amplitudes of the individual waves. Let us assume here that the component waves are of the *same amplitude*.

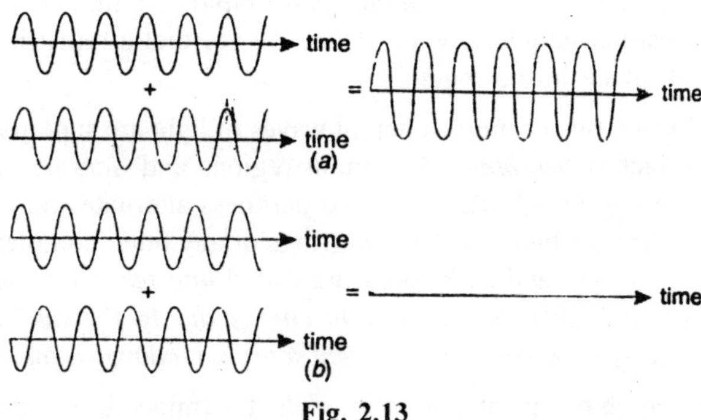

Fig. 2.13

At certain points, the two waves may be in *phase*. The amplitude of the resultant wave will then be equal to the sum of the amplitudes of the two waves, as shown in Fig. 2.13(a). Thus, the amplitude of the resultant wave

$$A_R = A + A = 2A. \quad ...(1)$$

Hence, the intensity of the resultant wave is

$$I_R \propto A_R^2 = 2^2 A^2A^2 = 2^2I. \quad ...(2)$$

It is obvious that the resultant intensity is greater than the sum of the intensities due to individual waves.

$$I_R > I + I = 2I \quad ...(3)$$

Therefore, the interference produced at these points is known as *constructive interference. A stationary bright band* of light is observed at points of constructive interference.

At certain other points, the two waves may be in *opposite phase.* The amplitude of the resultant wave will then be equal to the sum of the amplitudes of the two waves, as shown in Fig. 2.13 (b). Thus, the amplitude of the resultant wave

$$A_R = A - A = 0 \qquad ...(4)$$

Hence, the intensity of the resultant wave is

$$R_R \propto 0^2 = 0. \qquad ...(5)$$

It is obvious that the resultant intensity is less than the sum of the intensities due to individual waves.

$$I_R < 2I \qquad ...(6)$$

Therefore, the interference produced at these points is known as *destructive interference.* A *stationary dark band* of light is observed at points of destructive interference. Thus, we see that a redistribution of energy took place in the region.

Thus, when two or more coherent waves of light are superposed, the resultant effect is *brightness* in certain regions and *darkness* at other regions. The regions of brightness and darkness alternate and may take the form of straight bands, or circular rings or any other complex shape. The alternate bright and dark bands are called *interference fringes.* The *phenomenon of redistribution of light energy due to the superposition of light waves from two or more coherent sources is known as interference.*

Whether the occurs at a point is *solely* determined by the difference in the optical paths traversed by the waves that are superposing at that point.

Let us consider two sources of light S_1 and S_2 as shown in Fig. 2.14. Let us assume that the sources are identical and produce harmonic waves of same wavelength and that the waves are in the same phase at S_1 and S_2. Light from these sources travel along different paths, S_1P and S_2P, and meet at a point P. We now wish to know whether we get brightness or darkness at P due to the superposition of waves.

Referring to Fig. 2.14, we find that the waves move along the geometric paths $S_1P = r_1$ and $S_2P = r_2$, which are different in length. Also, the media through which the two waves travelled, may be different. As a result, the optical path lengths are different. If μ_1 is the refractive index of the medium in which the ray S_1P travelled, the corresponding optical path length is $\mu_1\, \mu_1$. Similarly, if μ_2 is the refractive index of the medium in which the ray S_2P travelled, the corresponding optical path length is $\mu_2 r_2$. These optical paths accommodate different number of waveforms along their lengths. The optical path difference between the waves at the point P is $(\mu_2 r_2 - \mu_1 r_1)$. It may come to a few full waves or a mixed fraction of waves. It means that though the waves started with the same phase, they may arrive at P with different phases because they travelled along different optical path lengths.

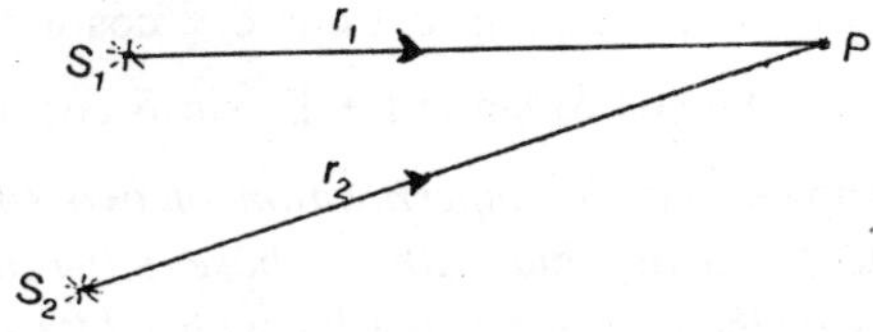

Fig. 2.14

If the optical path difference $\Delta = (\mu_2 r_2 - \mu_1 r_1)$ is equal to zero or an integral multiple of wavelength λ, then the waves arrive in phase at P and superpose with crest-to-crest correspondence. That is, if

$$\Delta = m\lambda \qquad ...(7)$$

where m is an integer and takes values, m = 0, 1, 2, 3, 4, 5,.......then the waves are in phase (Fig. 2.15a) and their overlapping at P produces constructive interference or brightness.

On the other hand, if the optical path difference $\Delta = (\mu_2 r_2 - \mu_1 r_1)$ is equal to an odd integral multiple of half-wavelength, $\lambda/2$, then the waves arrive out of phase at P and superpose with crest-to-trough correspondence. That is, if

$$\Delta = (2m + 1)\frac{\lambda}{2} \qquad ...(8)$$

where m is an integer and takes values, m = 0, 1, 2, 3, 4, 5,........then the waves are inverted with respect to each other and their overlapping at P produces destructive interference or darkness.

The regions of brightness and darkness are also known as regions of maxima and minima.

Theory of Interference

(a) *Analytical Method :* Let us assume that the electric field components of the two waves arriving at point P vary with time as

$$E_A = E_1 \sin \omega t \qquad ...(9)$$

and $E_B = E_2 \sin(\omega t + \delta)$ ...(10)

where δ is the phase difference between them. According to Young's principle of superposition, the resultant electric field at the point P due to the simultaneous action of the two waves is given by

$$E_R = E_A + E_B \quad ...(11)$$

$$= E_1 \sin \omega t + E_2 \sin(\omega t + \delta)$$

$$= E_1 \sin \omega t + E_2 (\sin \omega t \cos \delta + \cos \omega t \sin \delta)$$

$$= (E_1 + E_2 \cos \delta) \sin \omega t + E_2 \sin \delta \cos \omega t \quad ...(12)$$

Eqn. (12) shows that *the superposition of two sinusoidal waves having the same frequency but with a phase difference produces a sinusoidal wave with the same frequency but with a different amplitude E.*

Let $E_1 + E_2 \cos \delta = E \cos \phi$ ...(13)

and $E_2 \sin \delta = E \sin \phi$ ...(14)

where E is the amplitude of the resultant wave and ϕ is the new initial phase angle. In order to solve for E and ϕ, we square the eqn. (13) and (14) and add them.

$$(E_1 + E_2 \cos \delta)^2 + E_2^2 \sin^2 \delta = E^2 (\cos^2 \phi + \sin^2 \phi)$$

or $E^2 = E_1^2 + E_2^2 \cos^2 \delta + 2E_1E_2 \cos \delta + E_2 \sin^2 \delta$

or $E^2 = E_1^2 + E_2^2 + 2E_1E_2 \cos \delta$ ...(15)

Thus, it is seen that the square of the amplitude of the resultant wave is not a simple sum of the squares of the amplitudes of the superposing waves, there is an additional term which is known as the *interference term*.

(b) *Phasor Diagram and Phasor Addition :* A wave may be viewed either sideways or end on. In sideways view, as the wave travels through a distance λ, the phase angle changes from 0 to 2π radians. In the end-on view we find a point on the profile of the wave oscillating linearly. The two perspectives may be combined as follows. A circle having a radius OA equal to the amplitude of the wave motions is drawn (Fig. 2.15). Now consider a point P on the circumference of the circle; Q is the projection of P on the vertical axis. As P moves around the circumference with constant angular velocity ω, Q oscillates vertically. This is the end-on view. The Fig. 2.15 is called the *phasor diagram*. OP is called a *rotating vector* or a *phasor*. It means that the length of a phasor is proportional to the *amplitude* of the sinusoidal wave and the projection of a phasor on the vertical axis is proportional to the *instantaneous value* of the alternating quantity.

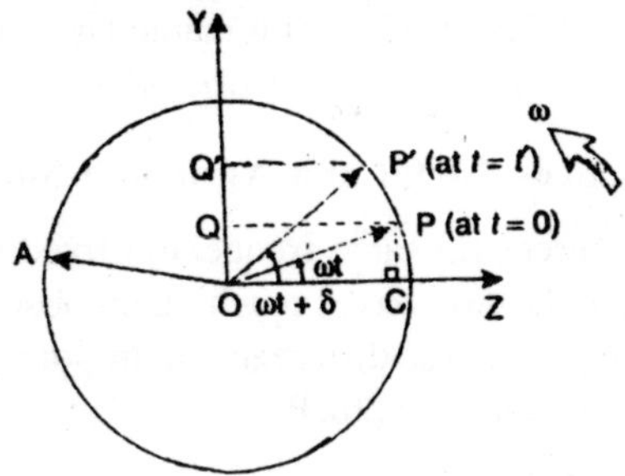

Fig. 2.15

Phasor representation may be used to *add sinusoidal functions* with a phase difference. The wave E_A is represented with a vector of amplitude E_1 rotating about the origin in a counter clockwise direction with an angular frequency ω (Fig. 2.16 a). As the phasor E_1 rotates, the projection E_A oscillates along the vertical axis. The second wave, E_B has amplitude E_2 and angular frequency ω but its phase is δ with respect to wave E_A. It is also shown in Fig. 2.16 (a). The resultant E_R is the sum of E_A and E_B obtained by drawing the phasors end to end, by placing the foot of one arrow at the head of the other (Fig. 2.16b), maintaining the proper phase difference. The whole assembly rotates counterclockwise about the origin. The sum of the projections on the vertical axis at any time gives the instantaneous value of the total field at a point. The amplitude E_R of the resultant sinusoidal wave at P is the vector sum of the other two phasors, as shown in Fig. 2.16 (b).

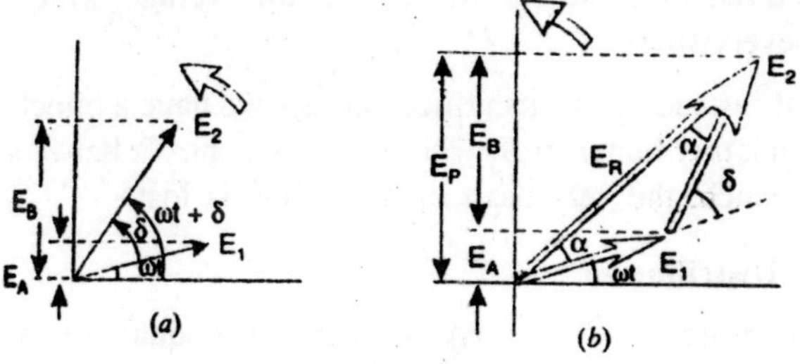

Fig. 2.16

To find E_R, we use the law of cosines.

$$E_R^2 = E_1^2 + E_2^2 - 2E_1E_2 \cos (\pi-\delta)$$

or $$E_R^2 = E_1^2 + E_2^2 + 2E_1E_2 \cos \delta$$

This is the same as the eqn. (21). This method is particularly convenient when several wave amplitudes have to be added.

Interference in Space—Nodal and Antinodal Surfaces

For a pair of sources (of same frequency) fixed in space, the phase difference δ between the two waves reaching an observation point P has two contributions: δ_0, the phase difference of the sources themselves, and δ_c due to the path difference upto P.

The locus of P for a given $\delta(= \delta_0 + \delta_c)$ is a surface in space; the locii for $\delta = (2n + l)n$ are called nodal surfaces, since node refers to a point of minimum oscillations; the locii for $\delta = 2n\pi$ are called antinodal surfaces. Thus we have the entire space filled with alternate nodal and antinodal surfaces, representing interference in space due to two sources.

The shapes of and spacings between these nodal and antinodal surfaces depend on the spacing of sources, the geometry, and A value.

If observations are limited to a plane (like a screen in the case of waves of light), what we observe is a cross-section of the nodal and antinodal surfaces—these will be nodal and antinodal lines. Notice that interference exists in all space; only our observations are limited to one plane, and we may shirt the observation plane in any way.

The role of δ_G is important to remember. If δ_0 changes by π. the nodal lines would become antinodal lines, and if δ_0 were rapidly varying with time we would not be able to observe the maxima and minimal. In principle, interference still exists in space, but the pattern is changing rapidly and randomly, so that we only see the average, which is uniform intensity everywhere

But, if instead of two identified sources we have a bunch of sources at S_1 and another bunch at S_2 with no phase controls between members of either bunch, the very basis of discussion is lost.

Intensity Distribution

The intensity of a light wave is given by the square of its amplitude.

$$I = 1/2\ \varepsilon_0 cE^2 \propto E^2$$

Using this relation into (15), we get

$$I = I_1 + I_2 + 2\sqrt{I_1 I_2}\ \cos\delta \qquad ...(16)$$

We see that the resultant intensity at P on the screen is not just the sum of the intensities due to the separate waves. The term $2\sqrt{I_1 I_2}\ \cos\delta$ is known as the *interference term.* Whenever the phase difference between the waves is zero, *i.e.* $\delta = 0$, we have maximum amount of light. Thus,

$$I_{max} = I_1 + I_2 + 2\sqrt{I_1 I_2} \qquad ...(17)$$

When $I_1 = I_2 = I_o \quad I_{max} = 4I_o \qquad ...(18)$

It means that the resultant intensity I will be *more than the sum* of the intensities due to the two sources.

When the phase difference is $\delta = 180°$, $\cos 180° = -1$ and we have a minimum amount of light.

$$I_{min} = I_1 + I_2 - 2\sqrt{I_1 I_2} \qquad ...(19)$$

which, when $I_1 = I_2$ becomes

$$I_{min} = 0 \qquad ...(20)$$

It means that the resultant intensity I will be *less than the sum* of the intensities due to the two sources.

At points that lie between the maxima and minima, when

$I_1 = I_2 = I_0$, we get

$$I = I_o + I_o + 2I_o \cos\delta$$

$$= 2I_o (1 + \cos\delta)$$

Then using the identity $1 + \cos\delta = 2\cos^2\left(\frac{1}{2}\delta\right)$, we get

$$I = 4I_o \cos^2\left(\frac{1}{2}\delta\right) \qquad ...(21)$$

Eqn. (21) shows that the intensity varies along the screen in accordance with the *law of cosine square.* Fig. 2.17 shows the variation of intensity as a function of phase angle δ.

It is seen from the plot that the intensity varies from zero at the fringe minima to $4I_o$ at the fringe maxima.

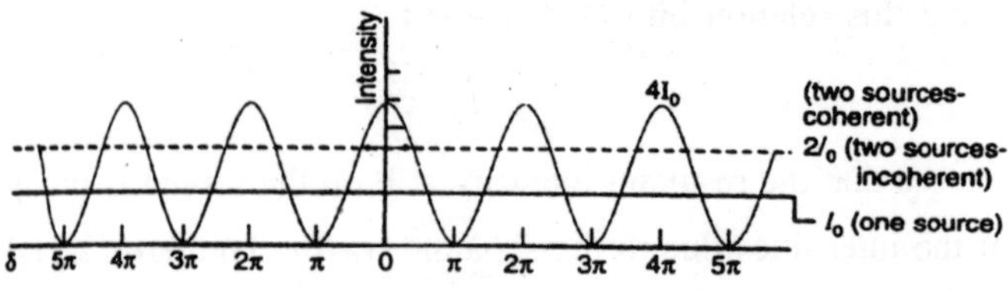

Fig. 2.17

Superposition of Incoherent Waves

Incoherent waves are the waves that do not maintain a constant phase difference. Then the phase of the waves fluctuate irregularly with time and independently of each other. In case of light waves the phase fluctuates randomly at a rate of about 10^8 per second. Light detectors such as human eye, photographic film etc. cannot respond to such rapid changes. The detected intensity is always the average intensity, averaged over a time interval which is very much larger than the time of fluctuation.

Thus, $$I_{are} = I_1 + I_2 + 2\sqrt{I_1 I_2}\ (\cos \delta)$$

The average value of the cosine over a large time interval will be zero and hence the interference term becomes zero. Therefore, the average intensity of the resultant wave is

$$I_{ave} = I_1 + I_2$$

If $I_1 = I_2$, then $I_{ave} = 2I$ ...(22)

It implies that the superposition of incoherent waves does not produce interference but gives a uniform illumination. The average intensity at any point is simply equal to the sum of the intensities of the component waves.

Superposition of Many Coherent Waves

The result (21) may be written as

$$I_{max} = 2^2\ I_0$$

which gives the resultant intensity when two coherent waves superpose. The resultant maximum intensity due to N coherent waves' will be therefore,

$$I_{max} = N^2\ I_o \qquad ...(23)$$

and the minimum intensity $I_{min} = 0$ ...(24)

where N represents the number of coherent waves superposing at a point.

APPLICATIONS OR MICHELSON INTERFEROMETER

Michelson interferometer can be used to determine :

(i) The wavelength of a given monochromatic source of light

(ii) The difference between the two neighbouring wavelengths or resolution of the spectral lines,

(iii) Refractive index and thickness of various thin transparent materials, and

(iv) For measurement of the standard metre in terms of the wavelength of light.

Measurement of Wavelength

Michelson interferometer is used to determine the wavelength of light from a monochromatic source. The monochromatic source is kept at S. If the mirrors M_1 and M_2 are exactly perpendicular, circular fringes are obtained. If the mirror M_1 is moved forward or backward, the circular fringes appear or disappear at the centre. Now, as the mirror is moved through a known distance d and the number of fringes disappearing at the centre is counted. Suppose d_1 is the initial thickness of the air film between the mirror M_1 and the image of M_2 corresponding to the bright fringe of order m_1 and d_2 is the final thickness of the air film corresponding to a bright fringe of order m in the same position. Then,

$$2d_1 = m_1\lambda$$

and $2d_2 = m_n\lambda$

By subtraction, we get $2(d_2 - d_1) = (m_n - m_1)\lambda$

$\therefore \quad 2d = N\lambda$ where $(d_2 - d_1) = d$ and $(m_n - m_1) = N$

$$\therefore \quad \lambda = \frac{2d}{N} \qquad ...(1)$$

Determination of the Difference in the Wavelength of two Waves

If a source of light consists of two wavelengths λ_1 and λ_2, which differ slightly, then two sets of fringes corresponding to the two wavelengths are produced in a Michelson interferometer. By adjusting the position of the mirror M_1 of the interferometer, the position is found when the fringes are very bright. In this position, the bright fringe due to λ_1 coincides with the bright fringes due to λ_2. When the mirror M_1 is moved, me two sets of fringes get out of step because their wavelengths are different. When the mirror M^ has been moved through a certain distance, the bright fringe due to one set will coincide with the dark fringe due to the other set and no fringes will be seen in this case. Again by moving the mirror M_1 a position is reached when a bright fringe of one set falls on the bright fringe of the other and the fringes are again distinct. This is possible when the mth order of the longer wavelength coincides with the $(m + 1)^{th}$ order of the shorter wavelength.

Let m_1 and m_2 be the changes in the order at the centre of the field when the mirror M_1 is displaced through a distance d between two consecutive positions of maximum distinctness of the fringes.

$$\therefore \qquad 2d = m_1 \lambda_1 = m_2 \lambda_2$$

If λ_1 is greater than λ_2,

$$m_2 = m_1 + 1$$

$$\therefore \qquad 2d = m_1\lambda_1 = (m_1 + 1)\lambda_2$$

$$\therefore \qquad m_1 = \frac{\lambda_2}{\lambda_1 - \lambda_2}$$

$$\therefore \qquad 2d = \frac{\lambda_1\lambda_2}{\lambda_1 - \lambda_2}$$

$$\text{or} \qquad \lambda_1 - \lambda_2 = \frac{\lambda_1\lambda_2}{2d}$$

Taking A as the mean wavelength of the two wavelengths λ_1 and λ_2, the small difference $\Delta\lambda$, is given by

$$\Delta\lambda = \lambda_1 - \lambda_2 = \frac{\lambda^2}{2d} \qquad ...(2)$$

Thickness of a Thin Transparent Sheet

Let a transparent sheet of thickness t and refractive index μ be inserted in the path of one of the interfering beams of Michelson interferometer. The optical path of that beam increases because of the sheet. It becomes μ t instead of t. The increase in the optical path is $(\mu t - 1)$ or $(\mu - 1)t$. Since the beam traverses the medium twice, the extra path difference between the two interfering beams is $2(\mu - 1)f$. If m is the number of fringes by which the fringe system is displaced, then

$$2 (\mu - 1) t = m\lambda$$

When monochromatic light is used, it is difficult to distinguish the sudden shift of fringes when the thin sheet is inserted. It is also not possible to count the number of fringes shifted. The difficulty is overcome by using white light first to locate the central dark fringe and it is made to coincide with the cross-wire of the telescope. The thin sheet is then introduced into the path of the beam. Position of mirror M_1 is adjusted till again a dark fringe of zero path difference coincides with the cross-wire of the telescope. The distance d through which the mirror is moved is noted. The white light is now replaced with the monochromatic light and the mirror M_1 is moved back slowly and the number of fringes contained in d is found. The thickness t is obtained from the relation

$$t = \frac{m\lambda}{2(\mu - 1)} \qquad ...(3)$$

Determination of the Refractive Index of Gases

When a tube containing a gas is introduced in the path of the beam going towards M_1, a path difference equal to $2(\mu - 1)\, l$ is introduced between the two interfering beams. Here, μ is the refractive index of the gas and l is the length of the tube. If m fringes cross the centre of the field of view, then $2(\mu - 1)\, l = m\lambda$. Knowing l, m, and λ, μ an be calculated.

In the path of the rays going towards M_1, a tube containing air at atmospheric pressure is introduced and the fringes are obtained in the centre of the field of view. In that case, refractive index of the air at various pressures can be determined. Let the length of the tube be l and let it contain air at atmospheric pressure. The tube is completely evacuated and m fringes cross the centre of the field of view. The path difference introduced between the two interfering beams is $2(\mu - 1)\, l$.

$$\therefore \qquad 2(\mu - 1)l = m\lambda \quad \therefore\ \mu = \frac{m\lambda}{2l} + 1$$

Standardisation of the Metre

The experiment to measure the standard metre in terms of the wavelength of the cadmium red line was first performed by Michelson and Benoit in 1895. It is not possible to count the number of fringes which cross the field of view when one of the mirrors of the Michelson interferometer is moved through whole length of one metre. Moreover, for a path difference of more than 20 cm, it is not possible to obtain the fringes. Therefore, the mirror must not be moved through a distance of more than 10cm. In practice nine etalons were used, each being twice the length of the preceding etalon. The length of the shortest etalon used is 0.390625 mm and of the longest was 10 cm. The experiment is divided into two main parts.

(i) The number of wavelengths of the monochromatic cadmium light is counted for the shortest etalon.

(ii) The length of the second etalon is compared with the shorter etalon and the process is repeated until the number of wavelengths for a length of 10 cm-etalon the known. From this 10 cm-etalon, the number of wavelengths for a length of one metre in terms of the wavelength of cadmium red line is known. This acts as a standard metre because, even if the original standard metre is destroyed, the standard metre can be formed again from the knowledge of the number of wavelengths. The standard metre is represented in terms of the wavelengths of red, green and blue lines of cadmium.

YOUNG'S DOUBLE SLIT EXPERIMENT—WAVEFRONT DIVISION

As early as in 1665 Grimaldi attempted to produce interference between two beams of light. He directed sunlight into a dark room through two pinholes in a screen, with an expectation that bright and dark bands would be observed in the area where the beams overlap on each other. He observed uniform illumination instead. In 1801, about one hundred thirty six years later, Thomas Young gave the first demonstration of the interference of light waves. Young admitted the sunlight through a single pinhole and then directed the emerging light onto two pinholes. Finally the light was received on a screen. The spherical waves emerging from the pinholes interfered with each other and a few coloured fringes were observed on the screen. The amount of light that emerged from the pinhole was very small and the fringes were faint and difficult to observe. The pinholes were later replaced with narrow slits that let through much more light. The sunlight was replaced by monochromatic light. Young's experiment is known as *double-slit experiment*.

Fig. 2.18 shows a plan view of the basic arrangement of the double slit experiment. The primary light source is a monochromatic source; it is generally a sodium lamp, which emits yellow light of wavelength at around 5893Å. This light is not suitable for causing interference because emissions from different parts of any ordinary source are *not* coherent. Therefore, the monochromatic light is allowed to pass through a narrow slit at S. The light coming out of the slit originated from only a small region of the light source and hence behaves more nearly like an ideal light source. Cylindrical wavefronts are produced from the slit S, the primary light source, which fall on the two narrow closely spaced slits, S_1 and S_2 as shown in Fig. 4.10. The slits at S_1 and S_2 are very narrow. The cylindrical waves emerging from the slits overlap. If the slits are *equidistant* from S, the phase of the wave at S_1 will be the same as the phase of the wave at S_2. Further, waves leaving S_1 and S_2 are therefore always *in phase*. Hence, sources S_1 and S_2 act as *secondary coherent sources*. The waves leaving from S_1 and S_2 interfere and produce alternate bright and dark bands on the screen at T.

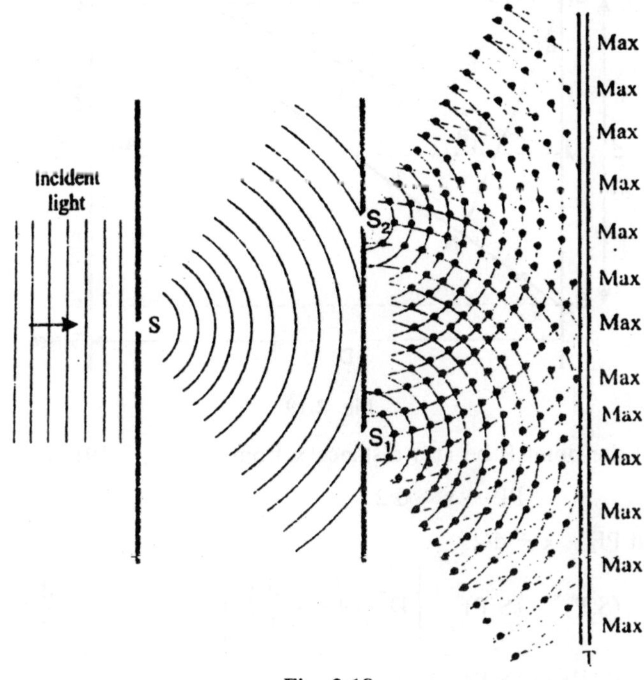

Fig. 2.18

Optical Path Difference Between the Waves at P

Let P be an arbitrary point on screen T, which is at a distance D from the double slits. Let θ be the angle between MP and the horizontal line MO. Let S_1N be a normal on to the line S_2P. The distances PS_1 and PN are equal. The waves emitted at the slits, S_1 and S_2 are initially in phase with each other? The difference in the path lengths of these two waves is S_2N. We assume that the experiment is carried out in air. Therefore, the optical paths are identical with geometrical paths. The nature of the interference of the two waves at P depends simply on how many waves are contained in the length of the path difference S_2N.

If S_2N contains an integral number of wavelengths, the two waves interfere constructively, producing a maximum in the intensity of light on the screen at P. If it contains an odd number of half-wavelengths, then the waves interfere destructively and produce a minimum intensity at P.

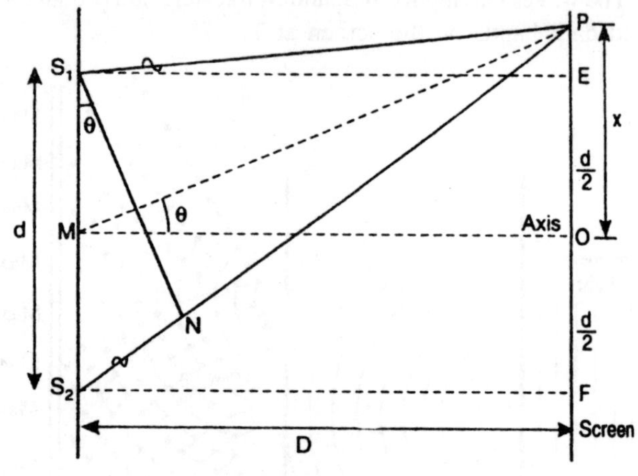

Fig. 2.19

Let the point P be at a distance x from O (Fig. 2.19). Then

$$PE = x - d/2$$

and $PF = x + d/2$.

$$(S_2P)^2 - (S_1P)^2 = \left[D^2 + \left(x + \frac{d}{2}\right)^2\right] - \left[D^2 + \left(x - \frac{d}{2}\right)^2\right]$$

$$(S_2P)^2 - (S_1P)^2 = 2xd$$

$$S_2P - S_1P = \frac{2xd}{S_2P + S_1P}$$

We can approximate that $S_2P \cong S_1P \cong D$.

$$\therefore \text{ Path difference} = S_2P - S_1P = \frac{xd}{D} \qquad ...(4)$$

We now find out the conditions for observing bright and dark fringes on the screen.

Bright Fringes

Bright fringes occur wherever the waves from S_1 and S_2 interfere constructively. The first time this occurs is at 0, the axial point. There, the waves from S_1 and S_2 travel the same optical path length to O and arrive in phase.

The next bright fringe occurs when the wave from S_2 travels one complete wavelength further the wave from S_1. In general constructive interference occurs if S_1P and S_2P differ by a whole number of wavelengths.

The condition for finding a bright fringe at P is that

$$S_2P - S_1P = m\lambda$$

Using the equation (29), it means that

$$\frac{xd}{D} = m\lambda \qquad ...(5)$$

where m is called the *order of the fringe.*

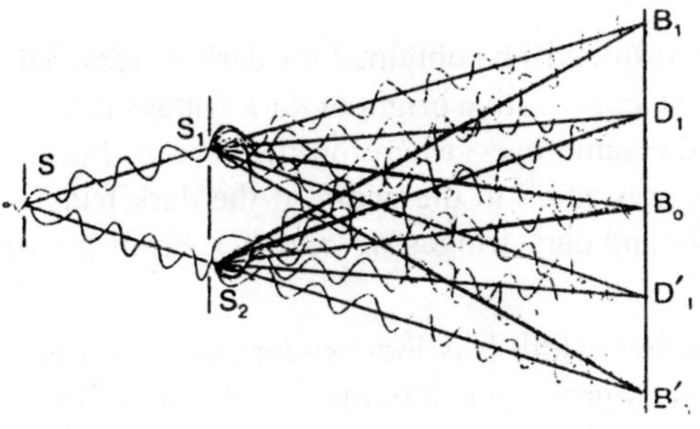

Fig. 2.20

The bright fringe B_0 (at 0), corresponding to m = 0, is called the *zero-order* fringe. It means the path difference between the two waves reaching at 0 is zero. Fringe at B_1 is the *first-order bright fringe* from the axis corresponding to m = 1; the path difference between the two waves reaching at B_1 is one λ. The *second order bright fringe* (m = 2) will be located where the path difference is 2X and so on.

Dark Fringes

The first dark fringe occurs when $(S_2P - S_1P)$ is equal to λ/1. The waves are now in opposite phase at P. The second dark ıringe occurs when $(S_2P - S_1P)$ equals 3λ/2. The mth dark fringe occurs when

$$(S_2P - S_1P) = (2m + 1)\lambda/2$$

The condition for finding a dark fringe is

$$\frac{xd}{D} = (2m+1)\frac{\lambda}{2} \qquad ...(6)$$

The *first-order dark fringe* D_1 (Fig. 2.20) from the axis corresponds to m = 0, where the path difference between the two waves is λ/2. The second order dark fringe (m = 1) will be produced where the path difference is 3λ/2 and so on.

Separation Between Neighbouring Bright Fringes

The mth order fringe occurs when $x_m = \frac{m\lambda D}{d}$

and the $(m + 1)^{th}$ order fringe occurs when $x_{m+1} = \frac{(m+1)\lambda D}{d}$

The fringe separation, β is given by $\beta = x_{m+1} - x_m = \frac{\lambda D}{d}$ ...(7)

The same result will be obtained for dark fringes. Thus, the distance between any two consecutive bright or dark fringes is known as the *fringe width* and is the same everywhere on the screen. Further, the width of the bright fringe is equal to the width of the dark fringe. Therefore, the alternate bright and dark fringes we *parallel*. From the eqn. (7), we find the following:

(i) The fringe width β is independent of the order of the fringe. It is directly proportional to the wavelength of light, *i.e.* $\beta \propto \lambda$. The fringes produced by red light are less closer compared to those produced by blue light.

(ii) The width of the fringe is *directly proportional* to the distance of the screen from the two slits, $\beta \propto D$. The farther the screen, the wider is the fringe separation.

(iii) The width of the fringe is *inversely proportional* to the distance between the two slits.

The closer are the slits, the wider will be the fringes.

COHERENCE

Interference fringes did not appear on the screen in the experiment of Grimaldi as he did not keep the slit S before the double slit arrangement. He obtained only a uniform illumination. It was so because the beams arriving at the screen were not coherent and the phase difference between them varied with time in a haphazard way. The reason for the lack of coherence lies in the very process of light emission. In ordinary sources of visible light, individual atoms are responsible for the emission of light. An atom leaving an excited state gives up the excess energy in the form of a burst of light (photon) and jumps to the lower normal state. The process of transition of the atom from an upper state to a lower state lasts for a brief time of about 10^{-8} sec. Therefore, the light emitted by an atom is not a continuous harmonic wave of infinite extension but is a wave train of finite length having a certain limited number of oscillations. It is impossible to say exactly when an atom may emit light because the emission is completely a random process.

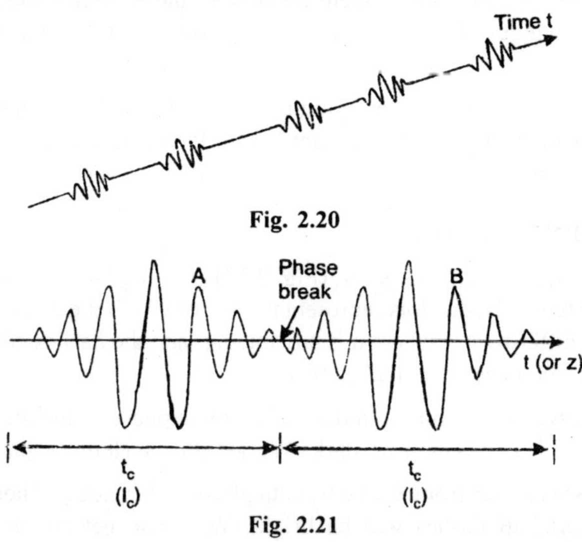

Fig. 2.20

Fig. 2.21

Fig. 2.20 pictorially describes the emission of light by a single atom in terms of wave trains. Other atoms in the source behave similarly but with different emission times. Adding together the wave trains generated by all atoms in the light source produces a succession of wave trains which have their phases distributed randomly. In passing from one wave train to the next there is an abrupt change in phase. Therefore, it is not possible to relate the phase at a point in wave train B to a point in wave train A. The phase of the wave train from an atom would remain constant with respect to the phase of the wave train from another atom only for about 10^{-8} sec. It implies that the two wave trains can be coherent for a maximum time of about 10^{-8} sec. Therefore, light from conventional sources is characterized by two important parameters, namely coherence time and coherence length.

Coherence Time : It is *the average time during which the wave remains sinusoidal and phase of the wave packet can be predicted reliably.*

Coherence Length : It is *the length of the wave packet over which it may be assumed to be sinusoidal and has predictable phase.* Light from a sodium discharge lamp has a coherence length of about 2 to 3 cm, while the coherence length of white light is a fraction of a cm. In the double slit experiment, the presence of slit S ensures that the same group of wave trains are incident on slits S_1 and S_2 When the phase of the wave changes at S this change is communicated simultaneously to S_1 and S_2.

Therefore, the waves emerging from S_1 and S_2 will be coherent with respect to each other and a stationary interference pattern is produced on the screen.

COHERENT SOURCES

Consider two sources S_1, S_2 (Fig. 2.22) emitting light of wavelength λ. On a screen AB, the dots represent points for which the distances from S_1 and S_2 differ by $n\lambda$, where n is any integer, and the dashes correspond to path differences $(n + 1/2)\lambda$. Now—

Let waves start from S_1 and S_2 in the same phase. Then all dots will be bright, all dashes will be dark. We get an interference pattern.

Let waves start from S_1 and S_2 with phase difference π. Then all dots will be dark, all dashes will be bright. We again get an interference pattern.

Let waves start from S_1 and S_2 with any other phase difference δ. We can again find a set of points which are bright and an alternate set of points which are dark.

So long as the sources S_1 and S_2 maintain any constant phase-relation, we always get a fixed interference pattern. But if the phase difference between S_1 and S_2 is changing it random with lime, the interference pattern will also change positions at random and we shall be observing only an average uniform intensity everywhere—no interference pattern.

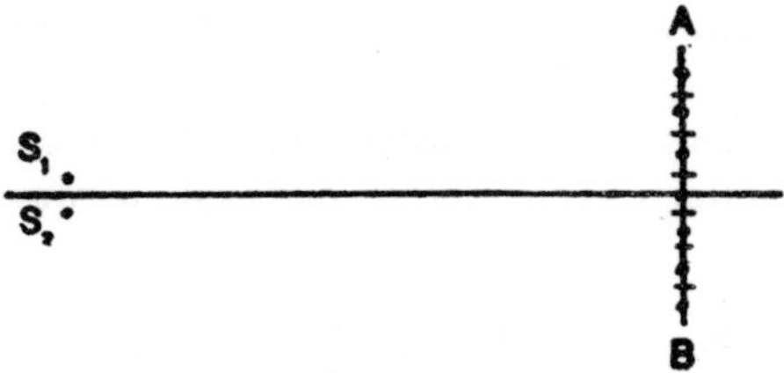

Fig. 2.22 : To explain the dependence of bright and dark positions on phase difference δ between the sources.

Two sources, such as S_1 and S_2 which maintain any constant phase relation between themselves are called coherent sources.

In the case of mechanical vibrations or electronic oscillators we can control two independent sources and maintain a constant phase relation between them. But the emission of light in any lamp comes from billions of atoms, the emission from each one of them being random. *Hence two independent sources of light can never be coherent.* In other words, interference of light can never be observed with two independent sources.

If *both sources of light are obtained from a single parent source by some device*, then all the random phase changes occurring in the parent source are repeated in the secondary ones also, so that relative phase remains constant. These are called coherent sources of light.)

CONDITIONS FOR INTERFERENCE

We may now summarize the conditions that are to be fulfilled in order to observe a distinct well-defined interference pattern.

(A) Conditions for Sustained Interference

(i) The waves from the two sources must be of the same frequency

If the light waves differ in frequency, the phase difference fluctuates irregularly with time. Consequently, the intensity at any point fluctuates with time and we will not observe steady interference.

(ii) The two light waves must be coherent

If the light waves are coherent, then they maintain a fixed phase difference over a time and space. Hence, a stationary interference pattern will be observed.

(iii) The path difference between the overlapping waves must be less than the coherence length of the waves.

We have already learnt that light is emitted in the form of wave trains and a finite coherence length characterizes them. If we consider two interfering wave trains, having constant phase difference, as in Fig. 2.23, the interference effects occur due to parts QR of wave 1 and ST of wave 2.

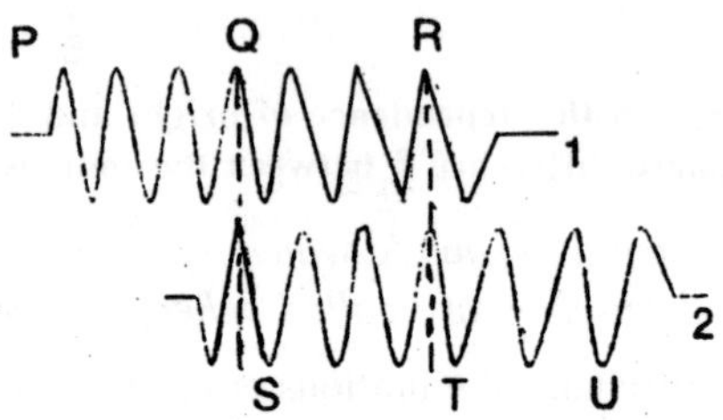

Partial overlap of the wavetrains

Fig. 2.23

For the parts PQ and TU interference will not occur. Therefore, the interference pattern does not appear distinctly. When the entire wave train PR overlaps on the wave train SU, interference pattern will be distinct, On the other hand, when the path difference between the waves 1 and 2 becomes very large, the wave trains arrive at different times and do not overlap on each other. Therefore, in such cases interference does not take place. The interference pattern completely vanishes if the path difference is equal to the coherence length. It is hence required that

$$\Delta < l_{coh} \qquad ...(1)$$

(iv) If the two sets of waves are plane polarized, their planes of polarization must be the same. Waves polarized in perpendicular planes cannot produce interference effects.

(B) Condition for Formation of Distinct Fringe Pattern

(v) The two coherent sources must lie close to each other in order to discern the fringe pattern. If the sources are far apart, the fringe width will be very small and fringes are not seen separately.

(vi) The distance of the screen from the two sources must be large.

(vii) The vector sum of the overlapping electric field vectors should be zero in the dark regions for obtaining distinct bright and dark fringes. The sum will be zero only if the vectors are anti-parallel and have the same magnitude.

TECHNIQUES OF OBTAINING INTERFERENCE

The phase relation between the waves emitted by two independent light sources rapidly changes with time and therefore they can *never* be coherent, though the sources are identical in all respects. However, if two sources are derived from a single source by some device, then any phase change occurring in one source is simultaneously accompanied by the same phase change in the other source. Therefore, the phase difference between the waves emerging from the two sources remains constant and the sources are *coherent*. The techniques used for creating coherent sources of light can be divided into the following two broad classes.

(a) *Wavefront Splitting :* One of the methods consists in dividing a light wavefront, emerging from a narrow slit, by passing it through two slits closely spaced side by side. The two parts of the same wavefront travel through different paths and reunite on a screen to produce fringe pattern. This is known as *interference due to division of wavefront*. This method is useful only with *narrow* sources. Young's double slit, Fresnel's double mirror, Fresnel's biprism, Lloyd's mirror, etc., employ this technique.

(b) *Amplitude Splitting :* Alternately, the amplitude (intensity) of a light wave is divided into two parts, namely reflected and transmitted components, by partial reflection at a surface. The two parts travel through different paths and reunite to produce interference fringes. This is known as *interference due to division of amplitude*. Optical elements such as beam splitters, mirrors are used for achieving amplitude division. Interference in thin films (wedge, Newton's rings etc.), Michelson's interferometer etc. interferometers utilize this method. This method requires *extended* source.

FRESNEL BIPRISM

Fresnel used a biprism to show interference phenomenon. The biprism consists of two prisms of very small refracting angles joined base to base. In practice, a thin glass plate is taken and one of its faces is ground and polished till a prism (Fig. 2.24 a) is formed with an obtuse angle of about 179° and two side angles of the order of 30'.

When a light ray is incident on an ordinary prism, the ray is bent through an angle called the *angle of deviation*. As a result, the ray emerging out of the prism appears to have emanated from a source S' located at a small distance above the real source, as shown in Fig. 2.24(b).

We say that the prism produced a *virtual image* of the source. A biprism, in the same way, creates two virtual sources S_1 and S_2 as seen in Fig. 2.24(c). These two virtual sources are images of the same source S produced by refraction and are hence *coherent.*

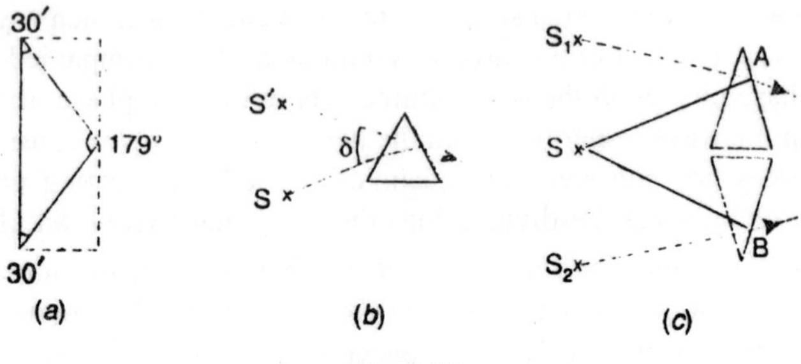

Fig. 2.24

Experimental Arrangement

The biprism is mounted suitably on an optical bench. An optical bench consists of two horizontal long rods, which are kept strictly parallel to each other and at the same level.

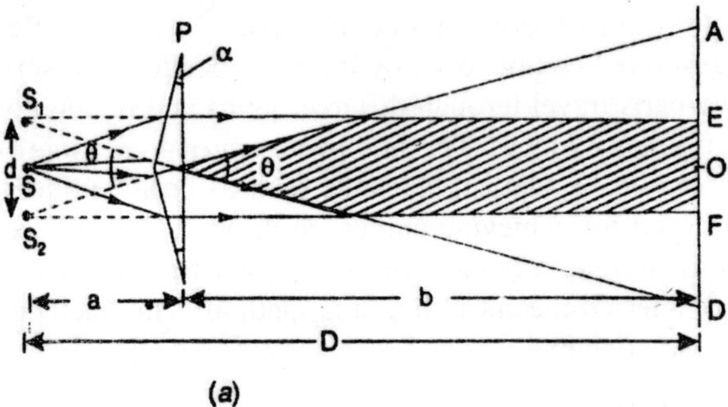

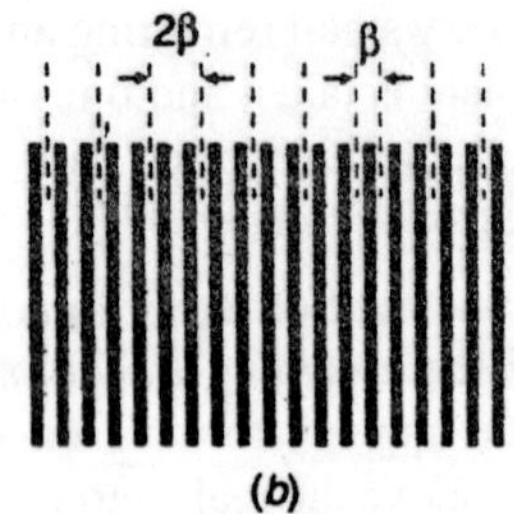

Fig. 2.25

The rods carry uprights on which the optical components are positioned. A monochromatic light source such as sodium vapour lamp illuminates a vertical slit S. Therefore, the slit S acts as a narrow linear monochromatic light source. The biprism is placed in such a way that its refracting edge is parallel to the length of the slit S. A single cylindrical wavefront impinges on both prisms. The top portion of wavefront is refracted downward and appears to have emanated from the virtual image S_1. The lower segment, falling on the lower part of the biprism, is refracted upward and appears to have emanated from the virtual source S_2. The virtual sources S_1 and S_2 are coherent (Fig. 2.25), and hence the light waves are in a position to interfere in the region beyond the biprism. If a screen is held there, interference fringes are seen. In order to observe fringes, a micrometer eyepiece is used.

Theory

The theory of the interference and fringe formation in case of Fresnel biprism is the same as described for the double-slit. As the point O is equidistant from S_1 and S_2 the central bright fringe of maximum intensity occurs there. On both sides of O, alternate bright and dark fringes, as shown in Fig. 2.25(b), are produced. The width of the dark or bright fringe is given by eqn. (31).

$$\beta = \frac{\lambda D}{d}$$

where D(= a + b) is the distance of the sources from the eye-piece.

Determination of Wavelength of Light

The wavelength of the light can be determined. For using the relation, the values of β, D and d are to be measured. These measurements are done as follows.

Adjustments

A narrow adjustable slit S, the biprism, and a micrometer eyepiece are mounted on the uprights and are adjusted to be at the same height and in a straight line. The slit is made vertical and parallel to the refracting edge of the biprism by rotating it in its own plane. It is illuminated with the light from the monochromatic source. The biprism is moved along the optical bench till, on looking through it along the axis of the optical bench, two equally bright vertical slit images are seen. Then the eyepiece is moved till the fringes appear in the focal plane of the eyepiece.

(i) *Determination of fringe width β:* When the fringes are observed in the field view of the eyepiece, the vertical cross-wire is made to coincide with the centre of one of the bright fringes. The position of the eyepiece is read on the scale, say x_o. The micrometer screw of the eyepiece is moved slowly and the number of the bright fringes N that pass across the cross-wire is counted.

The position of the cross-wire is again read, say x_N The fringe width is then given by $\beta = \frac{x_N - x_o}{N}$

(ii) *Determination of 'd'* : (a) A convex lens of short focal length is placed between the slit and the eyepiece without disturbing their positions. The lens is moved back and forth near the biprism till a sharp pair of images of the slit is obtained in the field view of the eyepiece. The distance between the images is measured. Let it be denoted by d_1.

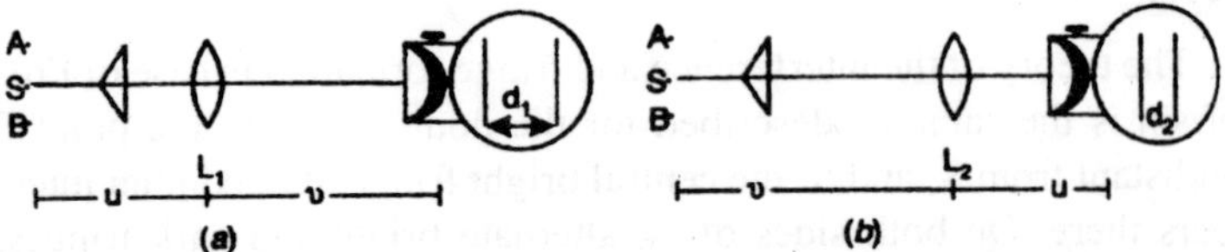

Fig. 2.26 : Measurement of the distance between the two virtual sources.

If u is the distance of the slit and u that of the eyepiece from the lens (Fig. 2.26a.), then the magnification is

$$\frac{\upsilon}{u} = \frac{d_1}{d} \qquad ...(1)$$

The lens is then moved to a position nearer to the eyepiece, where again a pair of images of the slit is seen. The distance between the two sharp images is again measured. Let it be d_2. Again magnification is given by

$$\frac{u}{\upsilon} = \frac{d_2}{d} \qquad ...(2)$$

Note that the magnification in one position is the reciprocal of the magnification in the other position.

Multiplying the equations (1) and (2), we obtain

$$\frac{d_1 d_2}{d^2} = 1$$

$$d = \sqrt{d_1 d_2} \qquad ...(3)$$

Using the values of β, d and D in the equation (31), the wavelength λ can be computed.

(b) Alternatively, the value of d can be determined as follows. The deviation δ produced in the path of a ray by a thin prism is given by

$$\delta = (\mu - 1)\alpha$$

where a is the refracting angle of the prism. From the Fig. 2.26, it is seen that ° = 9/2. Since d is very small, we can also write d = a θ.

$$\therefore \quad \frac{\theta}{2} = \frac{d}{2a} = (\mu - 1)\alpha$$

$$\therefore \quad d = 2a\,(\mu - 1)\,\alpha \qquad ..(4)$$

Interference Fringes with White Light

In the biprism experiment if the slit is illuminated by white light, the interference pattern consists of a central *white fringe* flanked on its both sides by a few coloured fringes and general illumination beyond the fringes. The central white fringe is the *zero-order* fringe.

With monochromatic light all the bright fringes are of the same colour and it is not possible to locate the zero-order fringe. Therefore, in order to locate the zero order fringe the biprism is to be illuminated by white light.

Lateral Displacement of Fringes

The biprism experiment can be used to determine the thickness of a given thin sheet of transparent material such as glass or mica. If a thin transparent sheet is introduced in the path of one of the two interfering beams, the fringe system gets displaced towards the beam in whose path the sheet is introduced. By measuring the amount of displacement, the thickness of the sheet can be determined.

Suppose S_1 and S_2 are the virtual coherent monochromatic sources. The point O is equidistant from S_1 and S_2, where we obtain the *central bright, fringe*. Therefore, the optical path $S_1O = S_2O$. Let a transparent plate G of thickness t and refractive index p, be introduced in the path of one of the beams (Fig. 2.27).

The optical path lengths S_1O and S_2O are now not equal and the *central bright fringe* shifts to P from O. The light waves from S_1 to P travel partly in air and partly in the sheet G; the distance travelled in air is $(S_1P - t)$ and that in the sheet is t.

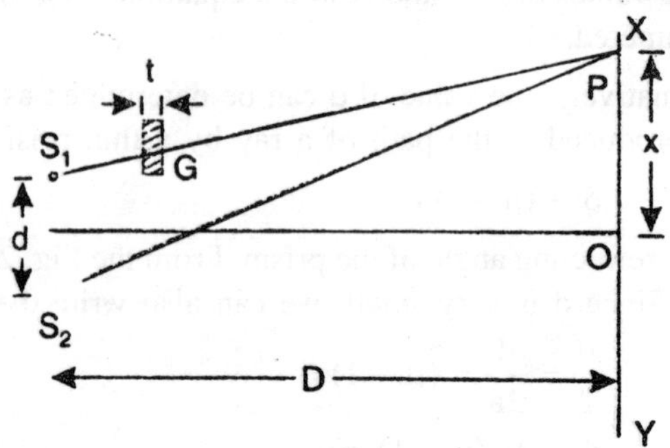

Fig. 2.27

The optical path $\Delta_{S_1P} = (S_1P - t) + \mu t = S_1P + (\mu - 1) t$

The optical path $\Delta_{S_2P} = S_2P$

The optical path difference at P is $\Delta_{S_1P} - \Delta_{S_2P} = 0$, since in the presence of the thin sheet, the optical path lengths S_1P and S_2P are *equal* and central zero fringe is obtained at P.

$$\therefore \qquad \Delta_{S_1P} = \Delta_{S_2P}$$

$$[S_1P + (\mu - 1)t] = S_2P$$

$$\therefore \qquad S_2P - S_1P = (\mu - 1) t$$

But according to the relation (28), $S_2P - S_1P = \dfrac{xd}{D}$ where x is the *lateral shift* of the central fringe due to the introduction of the thin sheet.

$$\therefore \qquad (\mu - 1) t = \frac{xd}{D}$$

Hence, the thickness of the sheet is $t = \dfrac{xd}{D(\mu - 1)}$ ... (5)

LLOYD'S SINGLE MIRROR

In 1834, Lloyd devised an interesting method of producing interference, using a single mirror and using almost grazing incidence. The Lloyd's mirror consists of a plane mirror about 30 cm in length and 6 to 8 cm in breadth (see Fig. 2.28). It is polished on the front surface and blackened at the back to avoid multiple reflections. A cylindrical wavefront coming from a narrow slit S_1 falls on the mirror which reflects a portion of the incident wavefront, giving rise to a virtual image of the slit S_2. Another portion of the wavefront proceeds directly from the slit S_1 to the screen. The slits S_1 and S_2 act as two coherent sources. Interference between direct and reflected waves occurs within the region of overlapping of the two beams and fringes are produced on the screen placed at a distance D from S_1 in the shaded portion EF.

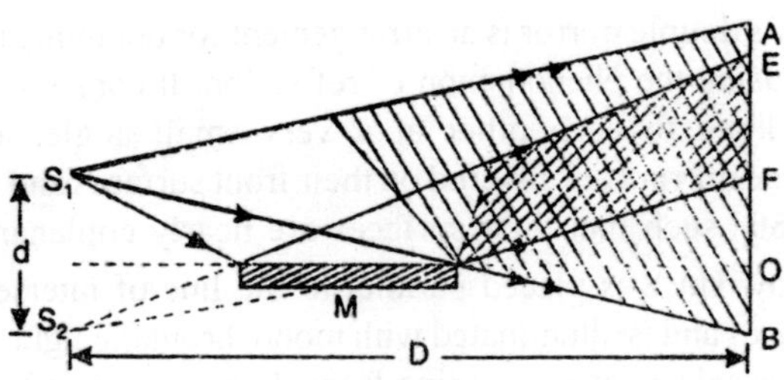

Fig. 2.28

The point O is equidistant from S_1 and S_2 Therefore, central (zero-order) fringe is expected to lie at O (the perpendicular bisector of S_1S_2) and it is also expected to be bright. However it is not usually seen since the point O lies outside the region of interference (only the direct light and not the reflected light reaches O).

By moving the screen nearer to the mirror such that it comes into contact with the mirror, the point O can be just brought into the region of interference.

With white light the central fringe at O is expected to be white but in practice it is dark. The occurrence of dark fringe can be understood taking into the consideration of the phase change of π that light suffers when reflected from the mirror. The phase change leads to a path difference of $\pi/2$ and hence destructive interference occurs there.

Determination of Wavelength

The fringe width. Thus,

Measuring β, D and d, the wavelength A, can be determined.

Comparison between the fringes produced by biprism and Lloyd's mirror:

1. In biprism the complete set of fringes is obtained. In Lloyd's mirror a few fringes on one side of the central fringe are observed, the central fringe being itself invisible.

2. In biprism the central fringe is bright whereas in case of Lloyd's mirror, it is dark.
3. The central fringe is less sharp in biprism than that in Lloyd's mirror.

FRESNEL'S DOUBLE MIRROR

Fresnel's double mirror is an arrangement for obtaining two coherent sources by using the phenomenon of reflection. It consists of two plane mirrors inclined to each other at a very small angle, as shown in Fig. 2.29. The mirrors are silvered on their front surfaces and are arranged at nearly 180° such that their surfaces are nearly coplanar.

A narrow slit S is placed parallel to the line of intersection of the mirror surfaces and is illuminated with monochromatic light. One portion of the cylindrical wavefront coming from slit S is reflected from the first mirror and another portion of the wavefront is reflected from the second mirror. After reflection, the light appears to diverge from S_1 and S_2 which are the virtual images of S. As the images S_1 and S_2 of the slit are derived from the same source S, they behave as two coherent sources, placed at a distance d apart. The waves diverging from S_1 and S_2 overlap and interference fringes are produced in the overlapping region EF on the screen. The fringes are of equal width.

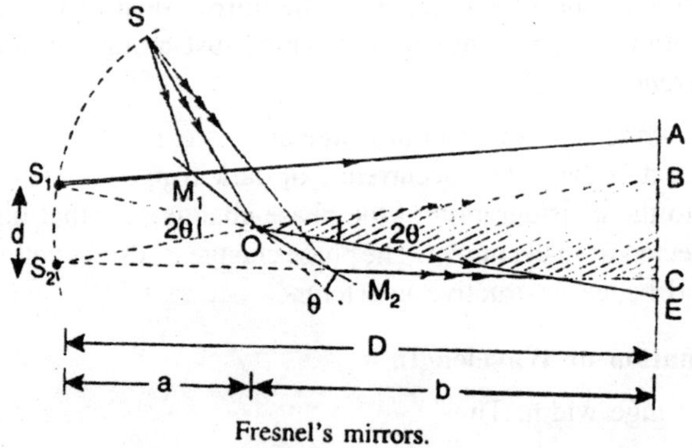

Fresnel's mirrors.

Fig. 2.29

Fringe Width

It is seen from the geometry of the (Fig. 2.29) that $OS_1 = OS_2 = OS$. That is, S_1, S_2 and S lie on a circle with O as a centre. Let 'a' be the distance of the sources and 'b' be the distance of the screen from O. Then the fringe width is given by

$$\beta = \frac{\lambda D}{d} = \frac{(a+b)\lambda}{d} \qquad ...(1)$$

OE and OB are the reflected rays from OM_1 and OM_2 respectively, corresponding to the incident ray SO. Therefore, the angle between OE and OB is twice the angle between the mirrors. Hence,

$$\angle S_1OS_2 = \angle BOE = 2\theta. \text{ Now, Arc } S_1S_2 = a \times 2\theta$$

$$\therefore \quad d = a \times 2\theta$$

Using the above result into eqn. (1), we get

$$\beta = \frac{(a+b)}{2a\theta}\lambda \qquad ...(2)$$

Comparison Between the Fringes Produced by Biprism and Double Mirror

The fringes in both cases are similar in appearance. However, the double mirror fringes are narrower than the biprism fringes.

ACHROMATIC FRINGES

A system of white and dark fringes, without any colours, obtained by white light is called *achromatic fringes*.

When the slit is illuminated by white light in any interference experiment, we obtain a central white fringe flanked by a few coloured fringes. Coloured fringes are obtained because the fringe width is dependent on the wavelength ($\beta = \lambda D/d$). For example, the width of the red fringe is more than the blue fringe. Rayleigh designed an experiment where white and dark fringes were obtained. It can be done if the fringe width is independent of the wavelength of light and is the same for all wavelengths. The fringe width β can be kept constant for all wavelengths, if λ/d is the same in all cases. Then the maxima of each order for all wavelengths coincide, resulting in achromatic fringes.

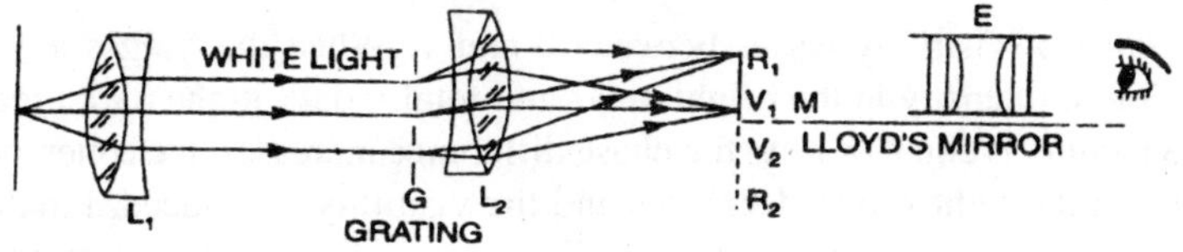

Fig. 2.30

In practice, achromatic fringes may be obtained as follows. S is a narrow source of white light at the focal plane of the converging lens L_1. A grating G having 800 to 1200 lines per cm is placed normal to light emerging from L_1. Another achromatic lens L_2 is used to form the second order spectrum on an opaque screen with a narrow opening in it. The narrow opening is adjusted so that only the first order spectrum is allowed to pass through it. The violet end is nearer to the highly polished Lloyd's nun-or M than the red end. The position of M is adjusted such that V_2 and R_2 are the images of V_1 and R_1. Interference occurs between the beams from V_1R_1 and those from V_2R_2. The violet fringes are produced by V_1 and V_2 while red fringes are produced by R_1 and R_2.

Suppose $V_1V_2 = d_1$ and $R_1R_2 = d_2$.

If $\frac{\lambda_V}{d_1} = \frac{\lambda_R}{d_2}$, the fringe width 3 will be the same and interference fringes due to different colours will overlap and white achromatic fringes are produced in the field of view. The white and dark fringes are seen through the eyepiece or can be projected on a screen.

NON-LOCALIZED FRINGES

Point sources produce fringes, which can be seen at different distances from the source. As the screen is moved far from the source, the fringe spacing increases and conversely, when the screen is moved nearer to the source, the fringes come closer. Therefore, we say the fringes are *non-localized.* Narrow sources produce non-localized fringes.

VISIBILITY OF FRINGES

The contrast of the interference fringes can be quantitatively described by the parameter called visibility. Visibility is defined as

$$V = \frac{I_{max} - I_{min}}{I_{max} + I_{min}} \quad ...(1)$$

the value of visibility varies between 0 and 1. When the fringes are of maximum intensity in the bright areas and totally dark in the dark areas, the visibility is equal to 1. As the phase difference increases, the coherence between the light waves decreases and the visibility is reduced. Finally, when the coherence between the two light waves disappears, I_{max} and I_{min} become equal and the visibility goes to zero. Then fringes are not observed and instead we observe uniform illumination.

The visibility of the fringes produced by two beams can be expressed as

$$V = \frac{(E_1 + E_2)^2 - (E_1 - E_2)^2}{(E_1 + E_2)^2 + (E_1 - E_2)^2} = \frac{2E_1E_2}{E_1^2 + E_2^2}$$

$$= \frac{2\sqrt{I_1 I_2}}{I_1 + I_2} = \frac{2\sqrt{I_1/I_2}}{1 + I_1/I_2}$$

It is seen from the above equation that the closer the intensities of the two waves, the higher is visibility of the fringes. When $I_1 = I_2$, $V = 1$. It may be noted that V is always equal to 1 when monochromatic light is used.

FRINGE PATTERN WITH WHITE LIGHT

When the slit is illuminated with white light in any interference experiment, we obtain a central white fringe flanked by a few coloured fringes. At the centre of the screen O, there is zero path difference and the bright fringes produced by all the colours there, add to each other. As a result a white fringe is produced at O. At other places away from O, the bright fringes in each order separate, because the fringe width is dependent on the wavelength ($\beta = \lambda D/d$). For example, the width of the red fringe is more than the blue fringe. Hence coloured fringes are produced on either side of the central white fringe.

INTERFEROMETRY

Instruments based on the principle of interference of light are known as *interferometers*. The instruments designed by Jamin and Rayleigh are used to determine the refractive index of gases and are known as *refractometers*.

Jamin's Refractometer

Jamin's refractometer consists of two exactly identical and optically plane glass blocks M_1 and M_2 (cut from the same block) and silvered on their back surfaces and arranged with their faces parallel to each other, as shown in Fig. 2.31. Light from an extended monochromatic source S incident on M_1 at an angle 45° is broken into two parallel rays (1) and (2) by reflection at the upper and lower surfaces of M_1. These two rays combine after suffering reflections at the two surfaces of M_2 as shown and form interference fringes as observed in a telescope. If the plates M_2 and M_2 are parallel, the light paths will be identical.

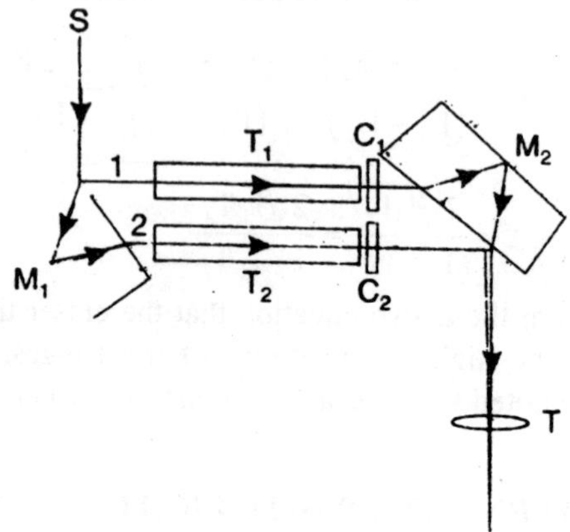

Fig. 2.31 : The Jamin interferometer.

Measurement of Refractive Index

To measure the refractive index of a gas at a given temperature and pressure, two similar evacuated tubes T_1 and T_2 are placed in the paths of the two parallel beams. The gas whose refractive index is to be measured is then slowly allowed to enter into one of the tubes. As the gas enters, the optical path of the beam passing through the tube increases. The fringes therefore move past the cross-wire of the telescope. The fringes are counted until the gas entering the tube attains the given pressure and temperature.

If l is the length of the tube and p. The refractive index of the gas, then the change in the optical path difference due to the presence of the gas is $(\mu - 1)\, l$. If N be the number of fringes passed across the field of view, then

$$(\mu - 1)\, l = Al \qquad ...(1)$$

The refractive index, μ can be calculated using the above expression.

Compensator

To avoid counting of fringes, a device called the "Jamin's compensator" is used. The compensator consists of two equally thick glass plates C_1 and C_2 cut from the same piece of glass and inclined at a small angle. The plates can be rotated together about a horizontal axis and the rotation is read on a divided circle D. One plate of the compensator is placed in the path of each beam. When the plates are equally inclined to the incident beam, the optical paths through the plates are the same. When the plates are rotated, the angles of incidence of the two beams change and a relative phase difference is introduced which varies as the compensator is rotated. By using monochromatic light and observing the passage of fringes across the field of view as the compensator is rotated, the scale; can be calibrated to read the optical path difference in terms of wavelength.

Now, in the actual experiment, white light is used and the central achromatic fringe is adjusted on the cross wire by adjusting the compensator. The gas is then introduced in one of the tubes at the given temperature and pressure which results in shifting of the fringes. The compensator is now rotated so as to bring the central fringe back on the cross wire again. The change in the optical path difference due to this rotation is determined as the compensator is already calibrated. This must be equal to $(\mu - 1)$ I, from which μ can be calculated.

Rayleigh Refractometer

The Rayleigh refractometer is mainly used to determine the refractive indices of inert gases and slight variation in the refractive index of solutions and gases. Light from a monochromatic source passes through a slit S and is then incident on the lens L_1.

The parallel beam then passes through the slits S_1 and S_2. The upper parts of the parallel beams emerging from S_1 and S_2 are allowed through the separate chambers T_1 and T_2. After passing through the chambers T_1 and T_2 and the compensating plates C_1 and C_2, the beams are recombined by the lens L_2 in its focal plane. Interference fringes are obtained in the focal plane of lens and are observed with the help of the eyepiece or a telescope. Because of the gas chambers S_1 and S_2 are widely separated and fringes are closely spaced.

The lens L_2 forms a *second* system of interference pattern, which is due to the superposition of the lower parts of the two beams emerging from S_1 and S_2 and passing beneath the chambers T_1 and T_2. The upper edge of the lower fringe system is made to coincide with the lower edge of the upper system with the help of an inclined thick plate P held across the lower parts of the slits. The centres of the two sets of interference patterns coincide and the fringe spacing in the two sets is the same. The first fringe system shifts with the change of gas pressure etc. whereas the second fringe system remains *stationary*.

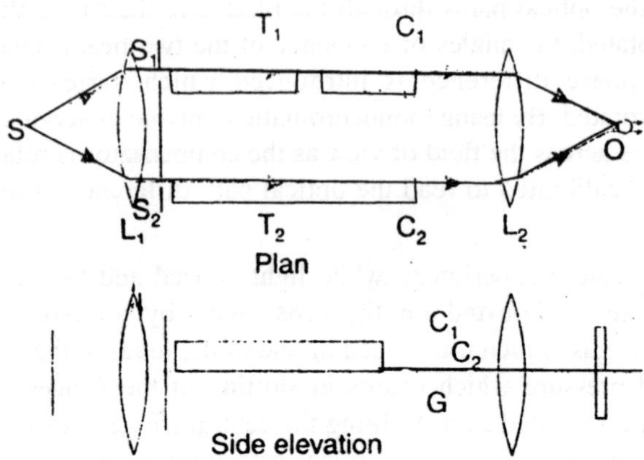

Fig. 2.32 : The Rayleigh interferometer.

The circular disc D attached to the compensating plates C_1 and C_2 is previously calibrated in terms of wavelength and refractive index. Initially, both the tubes T_1 and T_2 are evacuated. Using white light, the central white fringe is observed in the field of view of the eyepiece. The gas at a known pressure and temperature is introduced into the tube T_1. The central white fringe shifts from the field of view.

By rotating the circular disc D, which in turn displaces the compensating plates C_1 and C_2, the central white fringe is brought back to the centre of the field of view. The number of wavelengths graduated on the circular disc is noted. The change in the path difference is

$$(\mu - 1)l = N\lambda$$

where N is the number of fringes that crossed the field of view. Knowing l, N and λ, the refractive index p, can be calculated.

SOLVED EXAMPLES

Example 1:

The initial and final readings of a Michelson Interferometer screw are 10.7347 mm and 10.7051 mm as 100 fringes pass. Calculate the wavelength of light used.

Solution:

$$\Delta e = 10.7347 - 10.7051 = .0296 \text{ mm}$$

$$\lambda \frac{2\times.0296}{100} \text{ mm} = 5.92 \times 10^{-5} \text{ cm.}$$

Example 2:

An air-cell of thickness 3.00 cm is introduced in one path of a Michelson Interferometer, and fringes are obtained with mercury green light ($\lambda = 5.46 \times 10^{5}$ cm). Calculate the number of fringes that pass the field of view as the pressure in the cell is changed from 76.0 cm to 16.4cm [n of air at 76.0 cm pressure is 1.000293 and (n – 1) is proportional to pressure].

Solution:

$$(n - 1) \text{ at } 16.4 \text{ cm pressure} = 0.000293 \times \frac{16.4}{76.0} = 0.000063$$

$\therefore$ δn in changing pressure from 76.0 to 16.4 cm = 0.000230

$\therefore$ change in one way path = t.dn = 3 × 0.000230 = 0.000690 cm

$$\therefore \quad \text{number of fringes passing} = \frac{2\times0.000690}{5.46\times10^{-5}} = 25.3$$

(The converse of this is used for measuring the refractive index of gases).

Example 3:

In an experiment for determining the refractive index of gas using Michelson interferometer a shift of 140 fringes is observed, when all the gas is removed from the tube. If the wavelength of light used is 5400 Å and the length of the tube is 20 cm, calculate the refractive index of the gas.

Solution:

Given that n = 140, $\lambda = 5460 \times 10^{-10}$m, l = 20 cm = 0.2 m.

$$\mu = 1 + \left(\frac{n\lambda}{2l}\right) = 1 + \left[\frac{140\times5460\times10^{-10}\text{m}}{2\times0.2\text{m}}\right]$$

$$= 1.00019.$$

Example 4:

A glass microscope lens (μ = 1.50) is coated with magnesium fluoride (μ = 1.38) film to increase the transmission of normally incident yellow light (λ = 5800 Å).

Solution:

Given that mg = 1.50, mf = 1.38, l = 5800 Å.

$$\text{Minimum film thickness } t_{min} = \frac{\lambda}{4\mu_f} = \frac{5800}{4\times1.38}\text{Å} = 1050\ \text{Å}.$$

Example 5:

Newton's rings are observed between a convex lens and a plane plate. The diameters of xth and $x + 5^{th}$ rings ar 11.37 and 14.28 units. Deduce the diameters of (i) – 5th, (ii) $x + 14^{th}$ rings.

Solution:

Δ (D^2) is proportional to the difference of the 'order' number. Here we have:

$$D\ (D^2) \text{ for 5 rings} = 14.28^2 - 11.37^2$$

$$= 203.9 - 129.3 = 74.6\ (\text{units})^2$$

$$\therefore \quad D^2_{x-5} = 129.3 - 74.6 = 54.7;$$

$$\Rightarrow \quad D_{x-5} = 7.40 \text{ units}$$

$$\therefore \quad \Delta\ (D^2) \text{ for 9 rings} = 74.6 \times 9/5 = 134.3\ (\text{units})^2$$

$$\therefore \quad D_{x+14}^{\ 2} = 203.9 + 134.3 = 338.2$$

$$\Rightarrow \quad D_{x+14} = 18.40 \text{ units.}$$

Example 6:

A uniform plate of thickness 3.00 mm and n = 1.50 is viewed in light of λ = 6.00x 10^{5} cm. Calculate :

(i) order of interference m_0 at the centre, and

(ii) angular radius of fringe of order m_0 – 10.

Solution:

$$\text{(i) } m_0 = \frac{2en}{\lambda} = \frac{0.6\times1.50}{6.00\times10^{-5}} = 15{,}000$$

$$\text{(ii) } m_0 - 10 = \frac{2en}{\lambda}\left(1-\frac{\theta^2}{2}\right) = 15{,}000\left(1-\frac{\theta^2}{2}\right)$$

$$\therefore \quad \frac{\theta^2}{2} = \frac{\theta^2}{2}$$

$$\Rightarrow \theta = 0.037 \text{ rad.}$$

Example 7:

An equiconvex lens of focal length 80 cm (n of material 1.55) is placed over a plane plate. Newton's rings are formed with mercury green light (l = 5.46x 10 ⁵ cm). Deduce :

(i) Diameter of the first dar ring round the centre,

(ii) Difference of (Diameter)² for 10 fringes.

Solution.

For an equiconvex lens, $1/f = (n - 1). 2/R.$

That gives R = 88 cm

$$m = 1,$$

$$D_1 = [4 \times 88 \times 5.46 \times 10^{-5} \times 1]^{1/2} = 0.14 \text{ cm (check it)}$$

$$r = 10,$$

$$D_{m+10}^{\ 2} - D_m^{\ 2} = 4 \times 88 \times 10 \times 5.46 \times 10^{-5}$$

$$= 0.19 \text{ cm}^2 \text{ (check it).}$$

Example 8:

In a certain set-up the Newton's rings observed with light of λ = 5.89x 10 ⁵ cm show the difference of squares of diameters of successive rings as 0.124 cm². What happens to this quantity if we

(i) change l to 4.36x 10 ⁵ cm.

(ii) introduce a liquid of n = 1.56 between the lens and the plate,

(iii) the lens surface facing the plate is changed to one of twice the radius of curvature.

Solution:

The given quantity is

$$A = D_{m+1}^{\ 2} - D_m^{\ 2} = \frac{4R\lambda}{n} = 0.124 \text{ cm}^2$$

For case (i) $A'/A = \lambda'/\lambda$

$$\Rightarrow \quad A' = .124 \times \frac{4.36}{5.89} = 0.092 \text{ cm}^2$$

For case (ii) $A'/A = n'/n$

$$\Rightarrow \quad A' = .124 \times \frac{1}{1.56} = 0.079 \text{ cm}^2$$

For case (iii) A'/A = R'/R

$$\Rightarrow \quad A' = .124 \times \frac{2}{1} = 0.248 \text{ cm}^2.$$

Example 9:

Two waves of amplitudes 6 and 1 units are superposed with their vibrations parallel. Deduce the ratio of maximum to minimum intensity as the phase relation varies.

Solution:

Maximum amplitude = 6 + 1 = 7 units

Minimum amplitude = 6 – 1 = 5 units

$$\frac{I_{max}}{I_{min}} = \frac{7^2}{5^2} = \text{nearly } 2.$$

Example 10:

Five vibrations of equal amplitudes are superposed first with all in phase agreement, and then with successive phase differences 40°. Calculate the ratio of resultant intensities in the two cases.

Solution:

In the first case $\Sigma a_n \cos \delta_n = 5a$,

and $\Sigma a_n \sin \delta_n = 0$

$$\therefore \quad A_1 = [(5a)^2 + 0^2]^{1/2} = 5a$$

In the second case

$$\Sigma a_n \cos\delta n = a\,(1 + \cos 40° + \cos 80° + \cos 120° + \cos 160°) = 0.50a$$

$$\Sigma a_n \sin \delta_n = a\,(0 + \sin 40° + \sin 80° + \sin 120° + \sin 160°) = 2.84a$$

$$\therefore \quad A_2 = a\,[0.50^2 + (2.84)^2]^{1/2} = 2.88a$$

$$\therefore \quad \frac{I_1}{I_2} = \frac{A_1^2}{A_2^2} = \left[\frac{5a}{2.88a}\right]^2 = 3.0.$$

Example 11:

In a Michelson Interferometer set for white light a cover glass of thickness 2.31×10^{-3} cm and n = 1.53 is introduced in one path. If fringe width with Hg green light (λ = 5461 A) is 1.82×10^{-2} cm, by what distance would the white fringe shift?

Solution:

Additional optical path = $2(1.53 - 1) \times 2.31 \times 10^{-3}$ cm

$= 2.45 \times 10^{-3}$ cm

Equating this with $[y/(1.82 \times 10^{-2})] \times 5461 \times 10^{-3}$ cm gives

$$\text{shift } y = \frac{2.45\times10^{-3}1.82\times10^{-2}}{5.491\times10^{-5}} = 0.82 \text{ cm}$$

[Note that use of Eqn. (24a) would need measurement Δe which is $\sim 10^{-3}$ cm, whereas use of Eqn. (24b) needs measurement of y which is ~1cm -- in effect ~10^3 fold magnification].

Example 12:

Between two optically plane glass plates, at one edge a foilis introduced to get a wedge-shape air film. Viewed in mercury green light ($\lambda = 5.46 \times 10^{-5}$ cm) along the normal we see 12 fringes in 0.40 cm width. Deduce :

(i) the angle of the wedge;

(ii) thickness of the foil if plate length is 3.0 cm in all,

(iii) fringe width if water is introduced in the wedge space.

Solution:

Fringe-width w = (0.40/12) cm. Hence, we have

$$\theta = \frac{5.46\times10^{-5}}{2\times1\times0.40/12} = 8.2 \times 10^{-4} \text{ rad}$$

Now θ = foil thickness t ÷ plate length L. Hence

$$t = 8.2 \times 10^{-4} \times 3.0 = 2.5 \times 10^{-3} \text{ cm}$$

w is proportional to 1/n. Hence, with water,

$$w' = \frac{0.40}{12}\times\frac{1}{1.33} = 0.025 \text{ cm.}$$

Example 13:

A uniform soap film has thickness 3.0×10^{-5} cm. What colour does it show when seen in reflected white light along the normal?

Solution:

For soap film n = 1.33. The wavelengths for destructive interference in reflected light are given by.

$2 \times 1.33 \times 3.0 \times 10^{-5} = m\lambda$ (m = 1, 2...)

$\lambda = 8 \times 10^{-5}$ cm,

4×10^{-5} cm, 2.7×10^{-5} cm.

The possible λ values in the visible range are 8×10^{-5} cm (red) and 4×10^{-5} cm (violet). Hence these parts of the spectral colours are *absent* in the reflected light. Intermediate wavelengths correspond to yellow-green, and that colour will be seen.

Example 14:

If λ = = 5893 A° (Sodium D light), 2d = 1.00 mm, D = 50 cm, calculate the fringe-width.

Solution:

$$w = \frac{5893 \times 10^{-8} \times 50}{1 \times 10^{-1}} = 0.295 \text{ mm.}$$

Example 15:

In a particular two-slit interference pattern with λ = 6000 A° the zero-order and 10th order maxima fall at micrometer readings 12.34mm and 14.73 mm. If λ is changed to 5000 A° deduce the positions of the zero-order and 20th order fringes, other arrangements remaining the same.

Solution:

Zero-order fringe is independent of λ. So it falls at 12.34 mm.

Width with l = 6000 A is

$$w = \frac{14.73 - 12.34}{10} = 0.239 \text{ mm}$$

For λ = 5000 A°, we get the new fringe-width

$$w' = 0.239 \times \frac{5000}{6000} = 0.199 \text{ mm}$$

$$\Rightarrow \quad 20w' = 3.98 \text{ mm}$$

∴ Position of 20th order fringe with λ = 5000Å is

$12.34 \pm 3.98 = 16.32$ mm, 8.36 mm.

Example 16:

In a biprism experiment the fringe-width with λ = 5.89 × 10 ⁵ cm is 0.431 mm. On introducing a mica sheet in one path the central fringe shifts by 1.89 mm. Calculate the thickness of the sheet (n = 1.59).

Solution:

Shift 0.431 mm, corresponds to path diff. 5.89×10^{-5} cm

$$\therefore \quad ,,\ 1.89\ ,,\ ,,\ \frac{5.89 \times 1.89}{0.431} \times 10^{-5} \text{ cm}$$

Equating this with (n – 1)t, we get

$$t = \frac{5.89 \times 1.89}{0.59 \times 0.431} \times 10^{-5} \text{ cm} = 4.4 \times 10^{-4} \text{ cm.}$$

Example 17:

In a biprism experiment, the obtuse angle of the biprism is 178°0' and μ = 1.524. If the slit to biprims distance is 5.0 cm and biprism to screen is 65.0 cm, calculate the fringe-width for light of λ = 5893 A°.

Solution:

Angular deviation by a thin angle prism is $(\mu - 1)\,\theta$. For the biprims, we get

$$2d = S_1S_2 = a\,(\mu - 1)\,(\theta_1 + \theta_2)$$

$$= (5.0)\,(0.524)\left(2 \times \frac{\pi}{180}\right) = 0.091\text{cm}$$

Here $\theta_1 + \theta_2 = 180° - 178° = 2° = 2\pi/180$ radian

$$w = \frac{\lambda D}{2d} = \frac{5893 \times 10^{-8} \times (5 + 65)}{0.091} \text{cm} = 0.045 \text{ cm.}$$

Example 18:

A thin sheet of a transparent material (m = 1.60) is placed in the path of one of the interfering beams in a biprism experiment using sodium light, l = 5890 Å. The central fringe shifts to a position originally occupied by the 12th bright fringe. Calculate the thickness of the sheet

Solution:

The thickness of the sheet, $t = \dfrac{n\lambda}{\mu - 1}$

It is given that μ = 1.60, n = 12, λ = 5890 Å

$$\therefore \quad t = \frac{12 \times 5890 \times 10^{10}\,\text{m}}{1.60 - 1}$$

$$= 1.18 \times 10^{-5}\text{m} = 0.12\ \mu\text{m.}$$

Example 19:

In an interference pattern with two coherent sources the intensity variation is found to be 5% of the average intensity. Deduce the relative intensities of the interfering sources.

Solution:

Let the *amplitude* ratio of the sources be a : 1 (a > 1)

Then in the interference pattern the intensity ratio is

$(a + 1)^2 : (a - 1)$

In the present case,

$$\therefore \quad \frac{(a+1)^2}{(a-1)^2} = \frac{105}{95}, \Rightarrow a = 40 \text{ (check)}$$

$\therefore$ *Intensity* ratio of the sources = $a^2 : 1 = 1600 : 1$.

Example 20:

Green light of wavelength 5100 Å from a narrow slit is incident on a double slit. If the overall separation of 10 fringes on a screen 200 cm away is 2 cm, find the slit separation.

Solution:

The fringe width $\beta = \frac{\lambda D}{d}$

It is given that D = 200 cm,

$\lambda = 5100 \times 10^{-8}$cm

and $\beta = 2$ cm.

$\therefore \quad b = 0.2$ cm.

The slit separation $d = \frac{\lambda D}{d} = \frac{5100 \times 10^{-8}\,\text{cm} \times 200\,\text{cm}}{0.2\,\text{cm}} = 0.05$ cm.

Example 21:

In the above case give a plot of intensity versus phase difference.

Solution:

We have $I = 6^2 + 1^2 + 2 \times 6 \times 1 \cos\delta = 37 + 12 \cos\delta$.

The variation $12 \cos\delta$ superposed over a background of 39 units.

Example 22:

Two coherent sources are 0.18 mm apart and the fringes are observed on a screen 80 cm away. It is found that with a certain monochromatic source of light, the fourth bright fringe is situated at a distance of 10.8mm from the central fringe. Calculate the wavelength of light.

Solution:

The distance of the nth fringe from the central fringe is $x = \frac{n\lambda D}{d}$.

It is given that D = 80 cm, d = 0.18 mm = 0.018 cm,

x = 10.8 mm = 1.08 cm and n = 4.

$$\therefore \quad \lambda = \frac{xd}{nD} = \frac{1.08\text{cm} \times 0.018\text{cm}}{4 \times 80\text{cm}}$$

$$= 6075 \times 10^{-18} \text{ cm} = 6075\text{Å}.$$

Example 23:

Interference fringes are observed with a biprism of refracting angle 1° and refractive index 1.5 on a screen 80 cm away from it. If the distance between the source and the biprism is 20 cm, calculate the fringe width when the wavelength of light used is 6900Å.

Solution:

The fringe width $\beta = \frac{\lambda D}{d}$ and $d = 2(\mu - 1)\alpha\ a$

It is given that $\mu = 1.5$, $\alpha = 1° = \frac{\pi}{180°}$,

a = 20 cm and b = 80 cm, l = 6900 × 10^{-8}cm

$\therefore$ D = (20 + 80) cm = 100 cm.

$$\beta = \frac{\lambda D}{2(\mu - 1)\alpha\ a} = \frac{(6900 \times 10^{-8}) \times 100\text{cm}}{2(1.5 - 1) \times (\pi / 180) \times 20\text{cm}} = 0.02 \text{ cm}$$

Example 24:

In a biprism experiment the eyepiece is placed at a distance of 1.2m from the source. The distance between the virtual sources was found to be 7.5 × 10^{-4}m. Find the wavelength of light, if the eyepiece is to be moved transversely thorough a distance of 1.888 cm for 20 fringes.

Solution:

The fringe width $\beta = \frac{\lambda D}{d}$;

But $\beta = \frac{l}{n} \quad \therefore \quad \lambda = \frac{ld}{nD}$

It is given that l = 1.888 cm = 0.01888 m,

$d = 7.5 \times 10^{-4}$m, n = 20

and D = 1.2 m

$$\therefore \quad \lambda = \frac{0.01888\text{cm} \times 7.5 \times 10^{-4}\text{m}}{20 \times 1.2\text{m}} = 5900 \times 10^{-10}\text{m} = 5900 \text{ Å}.$$

Example 25:

When a thin sheet of transparent material of thickness 6.3 × 10^{-4} cm is introduced in the path of one of the interfering beams, the central fringe shifts to a position occupied by the sixth fringe. If λ = 5460 Å, find the refractive index of the sheet.

Solution:

$(\mu - 1)t = n\lambda$ ∴ The refractive index of the sheet $\mu = \frac{n\lambda}{t} + 1$.

It is given that $t = 6.3 \times 10^{-4}$cm, n = 6

and λ = 5460 Å = 5460 × 10–8cm.

$$\therefore \quad \mu = \frac{6 \times 5460 \times 10^{-8}\text{cm}}{6.3 \times 10^{-4}\text{cm}} + 1 = 1.52.$$

Example 26:

A light source emits light of two wavelengths 4300 Å and 5100 Å. The source is used in a double slit experiment. The distance between the sources and the screen is 1.5 m and the distance between the slits is 0.025 mm. Calculate the separation between the third order bright fringes due to these two wavelengths.

Solution:

It is given that $\lambda_1 = 4300$ Å $= 4300 \times 10^{-8}$cm,

$\lambda_2 = 5100$ Å $= 5100 \times 10^{-8}$cm, n = 3,

D = 1.5 m = 150 cm

and d 0.025 mm = 0.0025 cm.

Now $x_1 = \frac{n\lambda_1 D}{d}$ and $x_2 = \frac{n\lambda_2 D}{d}$

$$\therefore \quad x_2 - x_1 = \frac{nD}{d}(\lambda_2 - \lambda_1) \cdot$$

$$= \frac{3\times150\text{cm}}{0.025\text{cm}}(5100-4300)\times10^{-8}\text{cm} = 1.44 \text{ cm.}$$

Example 27:

In Lloyd's single mirror interference experiment, the slit source is at a distance of 2 mm from the plane of the mirror. The screen is kept at a distance of 1.5 m from the source. Calculate the fringe width. Wavelength of light is 5890 Å.

Solution:

The fringe width $b = \frac{\lambda D}{d}$

It is given that l = 5890 Å = 5890 × 10^{-10}m,

D = 1.5 m and d/2 = 2 mm. ∴ d = 4 × 10^{-3}m.

$$\beta = \frac{5890\times10^{-10}\times1.5\text{m}}{4\times10^{-3}\text{m}} = 22 \text{ mm.}$$

Example 28:

A soap film 5 × 10^{5} cm thick is viewed at an angle of 35° to the normal. Find the wavelengths of light in the visible spectrum which will be absent from the reflected light (μ = 1.33).

Solution:

Let i be the angle of incidence and r be the angle of refraction. Then

$$\mu = \frac{\sin i}{\sin r} \quad \therefore\ 1.33 = \frac{\sin 35°}{\sin r} \quad \therefore\ r = 25.5° \text{ and } \cos r = 0.90$$

The condition for destructive interference is $2\mu t \cos r = m\lambda$.

Using different values for m in the above relation, we get following values for wavelengths.

When m = 1, $\lambda_1 = 2 \times 1.33 \times 5 \times 10^{-5}\text{cm} \times 0.90$

$= 12.0 \times 10^{-5}\text{cm} = 120\ \mu\text{m}.$

When m = 2, $\lambda_2 = 2 \times 1.33 \times 5 \times 10^{-5}\text{cm} \times 0.90 \div 6.0 \times 10^{-5}\text{cm} = 6000$ Å.

When m = 3, $\lambda_3 = 2 \times 1.33 \times 5 \times 10^{-5}\text{cm} \times 0.90 \div 4.0 \times 10^{-5}\text{cm} = 4000$ Å.

When m = 4,

$\lambda_4 = 2 \times 1.33 \times 5 \times 10^{-5}\text{cm} \times 0.90 \div 3.0 \times 10^{-5}\text{cm} = 3000$ Å.

Out of the above wavelengths, $\lambda_2 = 6000$ Å, and $\lambda_3 = 4000$ Å lie in the visible region. Therefore, these two wavelengths are absent in the reflected light.

Example 29:

A glass wedge of angle 0.01 radian is illuminated by monochromatic light of wavelength 6000 Å falling normally on it, at what distance from the edge of the wedge will the 10th fringe be observed by reflected light?

Solution:

Given that $\theta = 0.01$ rad, $m = 10$,

$$\lambda = 6000 \times 10^{-8} \text{ cm}.$$

The condition for dark fringe is $2t = m\lambda$

The angle of the wedge $\theta = \frac{t}{x}$ or $t = \theta x$

$\therefore$ $$x\ \theta\ x = m\ \lambda$$

or $$x = \frac{\mu\lambda}{2\theta} = \frac{10 \times 6000 \times 10^{-8}\text{cm}}{2 \times 0.01} = 3 \text{ mm}.$$

Example 30:

A beam of monochromatic light of wavelength 5.82 × 10--7 m fall normally on a glass wedge with the wedge angle of 20 seconds of an arc. If the refractive index of glass is 1.5, find the number of dark fringes per cm of the wedge length.

Solution:

Given wedge angle $\theta = 20'' = \frac{20 \times \pi}{60 \times 60 \times 180}$ radians,.

$$\lambda = 5.82 \times 10^{-7} \text{ m}, \mu = 1.5.$$

Fringe width $$\beta = \frac{\lambda}{2\mu\theta} = \frac{5.82 \times 10^{-7}\text{ m} \times 60 \times 60 \times 180}{2 \times 1.5 \times 20 \times \pi} = 2 \text{ mm}.$$

Number of fringes per cm $= \frac{1}{0.2\text{cm}} = 5$ per cm.

Example 31:

Partially reflecting films can be coated on the surfaces of transparent media. In one case the coating reflects 80% intensity. Neglecting absorption, compute the relative amplitudes of successive beams in reflection and in transmission.

Solution:

Given $r^2 = 0.80$,

hence $t^2 = 1 - 0.80 = 0.20$

The successive reflected beams have amplitude ratios

1 : 0.20 : 0.16 : 0.13...

The transmitted beams have amplitude ratios

1 : 0.80 : 0.64 : 0.51 ...

Example 32:

A soap film is seen in sodium yellow light with rays falling at 30° from normal. Bright fringes occur at two points P and Q and there are 4 more bright fringes in between. Deduce the difference of thickness of the film between the points (n of soap film = 1.33).

Solution:

$$\sin i = 0.500, \therefore \sin r = \frac{0.500}{1.33} = 0.375$$

$$\Rightarrow \qquad \cos r = 0.927$$

Now we have, $\delta p = 5l$

$$\therefore \quad \delta e = \frac{\lambda p}{2n\cos r} = \frac{5\times 5.89\times 10^{-5}\,\text{cm}}{2\times 1.33\times 0.927} = 1.19 \times 10^{-4} \text{ cm.}$$

Example 33:

A wedge-shaped air-film between one fixed surface and another variable surface is seen with mercury green light ($\lambda = 5.46 \times 10^{-5}$ cm). An eyepiece is fixed on a given fringe, and as the variable surface moves perpendicular to itself we find 47 fringes passing across the cross-wire. Deduce the change in position of the surface.

Solution:

If initial fringe was of mth order, final fringe at the same place has $(m \pm 47)$ the order. Assuming that the fringes are seen nearly normally $(r = 0)$ an since the medium is air $(n = 1)$, we gives

$$2\Delta e = \pm 47\lambda = \pm 47 \times 5.46 \times 10^{-5} \text{ cm}$$

$\therefore \quad \Delta e = \pm 1.3 \times 10^{-3}$ cm.

(Note that this represent a typical use of interference fringes in measuring small displacements due to elastic bending, thermal expansion, etc.).

Example 34:

A thin equiconvex lens of focal length 4 m and refractive index 1.50 rests on and in contact with an optical flat, and using light of wavelength 5460 Å, Newton's rings are viewed normally by reflection. What is the diameter of the 5th bright ring?

Solution:

Given that m = 5, λ = 5460 Å = 5460×10^{-10}m,

$$f = 4\text{m}, \mu = 1.5$$

We know that $\frac{1}{f} = (\mu - 1)\left[\frac{1}{R_1} - \frac{1}{R_2}\right]$.

Here R_1 = and R_2 = – R

$$\therefore \quad \frac{1}{f} = (\mu - 1)\left[\frac{2}{R}\right]$$

$$\therefore \quad \frac{1}{4\text{m}} = 0.5\left[\frac{2}{R}\right] \text{ or } R = 4 \text{ m}.$$

The diameter of the mth bright ring is given by

$$D_m = \sqrt{2(2m-1)\lambda R}$$

$$= \sqrt{2(2\times5-1)\times5460\times10^{-10}\text{m}\times4\text{m}} = 6.2 \text{ mm}.$$

Example 35:

Newton's rings are observed in reflected light of l = 5.9 × 10 ⁵ cm. The diameter of the 10th dark rings is 0.5 cm. Find the radius of curvature of the lens and the thickness of the air film.

Solution:

Given that $\lambda = 5.9 \times 10^{-5}$cm, m = 10.

The radius of mth dark ring is given by

$$\therefore \quad R = \frac{r^2}{m\lambda} = \frac{(0.5\text{cm})^2}{10\times5.9\times10^{-5}\text{cm}} = 106 \text{ cm} = 1.06 \text{ m}$$

The thickness of air film is given by

$$\therefore \qquad t = \frac{m\lambda}{2} = \frac{10\times5.9\times10^{-5}\,\text{cm}}{2} = 2.95\ \mu\text{m}.$$

Example 36:

In a Newton's rings experiment, the diameter of 10th dark ring due to wavelength 6000 Å in air is 0.5 cm. Find the radius of curvature of the lens.

Solution:

$$\text{Radius of curvature, } R = \frac{(D/2)^2}{m\lambda}$$

$$= \frac{(0.5\times10^{-2}/2)^2\,\text{m}^2}{10\times6000\times10^{-10}\,\text{m}} = 10.4\ \text{m}.$$

Example 37:

In a Newton's rings experiment the diameter of the 15th ring was found to be 0.59 cm and that of the 5th ring was 0.336 cm. If the radius of the plano-convex lens is 100 cm, calculate the wavelength of light used.

Solution:

$$\lambda = \frac{D^2_{m+p} - D^2_m}{4pR} = \frac{D^2_{15} - D^2_5}{4\times10\times R}$$

$$= \frac{(5.9-3.36)^2\times10^{-6}\,\text{m}^2}{4\times10\times1\text{m}} = 5880\ \text{Å}.$$

Example 38:

In a Newton's rings experiment the diameter of 10th ring changes from 1.40 to 1.27 cm when a drop of liquid is introduced between the lens and the glass plate. Calculate the refractive index of the liquid.

Solution:

$$\mu = \frac{(D^2_m)_{air}}{(D^2_m)_{liq}} = \frac{(1.40\text{cm})^2}{(1.27\text{cm})^2} = 1.215.$$

Example 39:

In a Michelson interferometer 200 fringes cross the field of view when the movable mirror is moved through 0.0589 mm. Calculate the wavelength of light used.

Solution:

$$\lambda = \frac{2d}{m} = \frac{2\times0.0589\times10^{-3}\text{m}}{200} = 5890\ \text{Å}.$$

EXERCISES

1. Show that in interference with two coherent sources of any intensity ratio and phase different rf, the intensity variation can be expressed as either a cos J variation or a $\cos^2(\delta/2)$ variation.
2. Show that if n sources of equal intensity send out waves at a point with phase 0, ϕ, 2ϕ... $(n-1)\phi$, the resultant oscillation is zero if $n\phi$ is an integral multiple of 2π.
3. What is meant by interference of light? State the fundamental conditions for the production of interference fringes.
4. Two coherent sources, whose intensity ratio is 9:4, produce interference fringes. Deduce the ratio of maximum to minimum intensity of the fringe system.
5. What are coherent sources? Explain the importance of such sources in interference phenomenon. Two coherent sources form interference fringes. Obtain an expression for the distance between two consecutive bright fringes.
6. What are coherent sources? Discuss why two independent sources of light of the same wavelength cannot produce interference fringes? Give a diagram showing clearly how coherent sources are produced in a biprism. Derive the formula for the fringe width in the biprism experiment.
7. Interference fringes are formed by a biprism whose acute angle is 20' and refractive index is 1.5. The slit is at 10 cm from the biprism and is illuminated by light of wavelength 6000 Å. Find the fringe width on a screen placed at a distance of one metre from the biprism.
8. A biprism is placed 5 cm from a slit illuminated by sodium light (λ = 5890 Å). The width of the fringes obtained On a screen 75 cm from the biprism is 9.424×10^{-2} cm. What is the distance between the two coherent sources?
9. A biprism forms interference fringes with monochromatic light of wavelength 5450 Å. On introducing a thin glass plate (μ = 1.5) in the path of one of the interfering beams, the central

fringe shifts to the position previously occupied by the third bright fringe. Does fringe width change? Find the thickness of the plate.

10. State the basic conditions for the phenomenon of interference of light. Briefly discuss the effect of introducing a thin plate in the path of one of the interference beams in a biprism experiment. Deduce an expression for the displacement of the fringes. Show how this method is used for finding the thickness of a mica sheet.

11. Discuss the conditions for interference. Describe Young's experiment and derive an expression for :

    (i) intensity at a point on the screen, and

    (ii) fringe width.

12. Derive an expression for the resultant intensity when two coherent beams of light-are superposed.

    What is the visibility of fringes :

    (a) for two slits of equal intensities?

    (b) if intensity of one slit is 4 times the other?

    What will be the intensity when the two sources are incoherent?

13. A man standing 2.0 m above an oil film over water observes greenish colour at a point 2.5 m from his feet. Deduce the possible thickness of the oil film there.

    (λ_{green} = 5500Å, n_{oil} = 1.40, n_{water} = 1.33).

14. In an experiment to observe Newton's rings we may find that:

    (i) $D_{m+r}^2 - D_m^2$ does not come out as independent of m,

    (ii) the ringe are not quite circular,

    (iii) the centre is not dark, but is bright.

    What would you conclude in each case?

15. In a Newton's rings arrangement $D_{m+r}^2 - D_m^2$ came to 0.204cm^2 with air film. Deduce the $D_{p+r}^2 - D_p^2$ value (r same as before) if a drop of liquid of n = 1.473 is introduced as the film. Will the fringes (seen in reflection, as usual) be more intense with the liquid as film?

16. In what respects are Haidinger's fringes distinct in character from fringes in thin films?

17. Newton's rings are observed between a spherical, surface of radius of curvature 120 cm and a plane plate. The diameters of 5th and 15th bright rings are 0.314 cm and 0.584 cm. Calculate the diameters of 25th and 37th bright rings, and also the wavelength of light used.

18. Fringes in Michelson's interferometer and Fabry and Perot interferometer do not differ in character, but differ widely in intensity distribution. Comment.

19. Explain the working of Michelson interferometer. How will you produce circular fringes with it? How will you measure the difference in wavelengths between the D-lines of sodium light?

20. What are Newton's rings and how are they formed? How can the refractive index of a liquid be determined using these fringes? What is the difference between these fringes and those produced by a biprism?

21. Two microwave oscillators S_1, S_2 send out waves of $\lambda = 8.5$cm. Observations are made along the line S_1, S_2. Show that for fixed S_1, S_2 distance there are no alternate maxima-minima along this line. Also explain what happens if distance S_1, S_2 is varied from 10.0cm to 30.0 cm.

22. How can coherent sources be obtained in practice?

23. Is it necessary that the interfering waves should have the same frequency? If so, why?

24. Is it necessary that the interfering waves should have equal amplitudes? Explain.

25. Newton's rings are observed normally in reflected light of wavelength 5893Å. The diameter of the 10th dark ring is 0.005 m, find the radius of curvature of the lens and the thickness of the air film.

26. When the movable mirror of Michelson interferometer is shifted by 0.0589 mm, a shift of 200 fringes is observed. Find the wavelength of the light used.

27. Describe Michelson interferometer and explain the formation of fringes in it. How was this interferometer used for the standardization of the metre?

28. Discuss the principle and use of a Fabry-Perot interferometer.

29. Explain the formation of fringes in a Fabry-Perot interferometer and discuss the effect of reflectivity on the sharpness of fringes.
30. Describe the formation of fringes by Fabry-Perot interferometer and discuss the intensity distribution.
31. Explain with necessary theory the Newton's rings method of measuring the wavelength of light.
32. Discuss the formation of colours in thin transparent film due to multiple reflection of light in these and show that with monochromatic light the interference pattern of the reflected and the transmitted light are complimentary.
33. Show that the diameters of Newton's rings when tow surfaces of radii R_1 and R_2 are placed in contact, are related by the equation
$$\frac{1}{R_1} - \frac{1}{R_2} = \frac{4n\lambda}{d_n^2}$$
34. Describe the principle and method of production of interference filter.
35. A parallel beam of light (λ = 5890 Å) is incident on a thin glass plate (μ = 1.5) such that the angle of refraction into the plate is 60°. Calculate the smallest thickness of the glass plate, which will appear dark by reflection.
36. A soap film of refractive index 4/3 and of thickness 1.5×10^{-4} cm is illuminated by white light incident at an angle of 60°. the light reflected by it is examined by a spectroscope in which is found a dark band corresponding to a wavelength of 5×10^{-5}cm. Calculate the order of interference of the dark band.
37. Light of wavelength 6000Å falls normally on a thin wedge shaped film of refractive index 1.4, forming fringes that are 2mm apart. Find the angle of the wedge.
38. In an experiment with Michelson's interferometer the readings of two consecutive positions of the movable mirror for the maximum distinctness of fringes were found to be 1.2829 mm and 1.5774 mm. If the mean wavelength of sodium D-lines is 5893Å, find the difference between the two wavelengths.
39. Fringes of equal inclination are observed in a Michelson interferometer. As one of the mirrors is moved back by 1 mm, 3663 fringes move out from the centre of the pattern, Calculate the wavelength of light used.

40. Interference fringes in thin films are quite distinct when seen in reflection, but very indistinct in transmission. Explain. Will the statement hold good for films with reflecting coatings?
41. Explain why you need a wide source for seeing interference in films. Also explain why the observing instrument has to be focussed in a particular plane, unlike Young's fringes which are seen over a wide region of space.
42. A Fresnel biprism arrangement is set and in the field of view of the eyepiece we get δ_2 fringes with sodium light (λ = 5893Å). How many fringes shall we get in the same field of view if we replace the source with a mercury lamp using (a) green filter (λ = 5461Å), (b) violet filter (λ + 4358.4Å)?
43. Biprism fringes are produced with the following specifications: Source: sodium yellow (λ =5893 Å), biprism : μ = 1.50 and refracting angles 1.04° and 1.23°, distance from slit to biprism 12.4 cm and distance from slit to focal plane of the eyepiece 685 cm. Calculate :

    (a) the separation of the coherent slit images,

    (b) the fringe width.
44. In a biprism experiment fringes were first observed with sodium fight (A = 5893Å), and fringe-width was measured as 0347 mm. Then two thin transparent sheet A and B of thicknesses 0.016 mm and 0.020 mm and refractive indices 1.65 and 1.45 respectively were introduced in the two beams. Calculate the shift of the fringe system. Is the shift towards A or B?

3

Diffraction

INTRODUCTION

Fraunhoffer diffraction pattern, the incident wave front must be plane and the diffracted light is collected on the screen with the help of a lens. Thus, the source of light should either be at a large distance from the slit or a collimating lens must be used. Interference is seen easily for sound and for ripples on water. But in the case of light, even with the geometry set for very small wavelengths, interference is not observable until the special condition of coherence is satisfied. Hence, the earliest experimental evidence for wave-nature of light came from diffraction, which is defined as departure from the expectations of ray optics. In this chapter we will study these departures from the expectations of ray optics. The departures become observable when light passes through narrow slits or holes, or passes by narrow wires or obstacles.

FRAUNHOFFER DIFFRACTION AT DOUBLE SLIT

In Fig. 3.1, AB and CD are two rectangular slits parallel to one another and perpendicular to the plane of the paper. The width ,of each slit is a and the width of the opaque portion is b. L is a collecting lens and MN is a screen perpendicular to the plane of the paper. P is a point on the screen such that OP is perpendicular to the screen. Let a plane wave front be incident on the surface of XY. All the secondary waves traveling in a direction parallel to OP come to focus at P. Therefore, P corresponds to the position of the central bright maximum.

In this case, the diffraction pattern has to be considered in two parts (i) the interference phenomenon due to the secondary waves emanating from the corresponding points of the two slits and (ii) the diffraction pattern due to the secondary waves from the two slits individually.

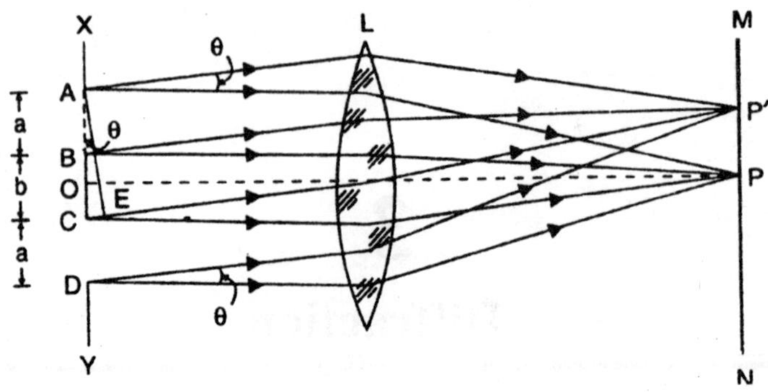

Fig. 3.1

For calculating the positions of interference maxima and minima, the diffracting angle is denoted as θ and for the diffraction maxima and minima it is denoted as ϕ. Both the angles θ and ϕ refer to the angle between the direction of the secondary waves and the initial direction of the incident light.

Distinction Between Single Slit and Double Slit Diffraction Patterns

The single slit diffraction pattern consists of a central bright maximum with secondary maxima and minima of gradually decreasing intensity. The double slit diffraction pattern consists of equally spaced interference maxima and minima with in the central maximum. The intensity of the central maximum in diffraction pattern due to a double slit is four times that of the central maximum in the diffraction pattern due to diffraction at a single slit. In the above arrangement, if one of the slits is covered with opaque screen, the pattern observed is similar to the one observed with a single slit. The spacing of diffraction maxima and minima depends on a, the width of the slit and the spacing of the interference maxima and minima depends on the value of a and b where b is opaque spacing between the two slits. The intensities of the interference maxima are not constant but decrease to zero on either side of the central maximum. These maxima reappear two or three times before the intensity becomes too low to be observed.

Missing Orders in a Double Slit Diffraction Pattern

In the diffraction pattern due to a double slit discussed earlier, the slit width is taken as a and the separation between the slits as b. If the slit width a is kept constant, the diffraction pattern remains the same. Keeping a constant, if the spacing b is altered the spacing between the interference maxima changes. Depending on the relative values of a and b certain orders of interference maxima will be missing in the resultant pattern.

The directions of interference maxima are given by the equation,

$$(a + b) \sin \theta = n\lambda \qquad \text{...(1)}$$

The direction of diffraction minima are given by the equation,

$$a \sin \theta = p\lambda \qquad \text{...(2)}$$

In equations (1) and (2) n and p are integers. If the value of a and b are such that both the equations are satisfied simultaneously for the same value of θ, then the positions of certain interference maxima correspond to the diffraction minima at the same position on the screen.

(i) Let $a = b$

Then $2a \sin \theta = n\lambda$

and $a \sin \theta = p\lambda$

$\therefore$ $n/p = 2$

or $n = 2\,p$

If $p = 1,2,3$ etc.,

then $n = 1, 4, 6$ etc.

Thus, the orders 2, 4, 6 etc of the interference maxima will be missing in the diffraction pattern. There will be three interference maxima in the central diffraction maximum.

(ii) If $2a = b$, then

$$3\,a \sin 0 = n\lambda$$

and $a \sin 0 = p\lambda$

$\therefore$ $n/p = 3$

or $n = 3p$

If $p = 1, 2, 3,$

etc, $n = 3, 6, 9,$ etc.

Thus the orders 3, 6, 9 etc of the interference maxima will be missing in the diffraction pattern. On both sides of the central maximum, the number of interference maxima is 2 and hence there will be five interference maxima in the central diffraction maximum. The position of the third interference maximum also corresponds to the first diffraction minimum

(iii) If $a + b = a$ *i.e.*, $b = 0$

The two slits join and all the orders of the interference maxima will be missing. The diffraction pattern observed on the screen is similar to that due to a single slit of width equal to 2a.

INTERFERENCE AND DIFFRACTION

It is clear from the double slit diffraction pattern that interference takes place between the secondary waves originating from the corresponding points of the two slits and also that the intensity of the interference maxima and minima is controlled by the amount of light reaching the screen due to diffraction at the individual slits. The resultant intensity at any point on the screen is obtained by multiplying the intensity function for the interference and the intensity function for the diffraction at the two slits. The values of the intensity functions are taken for the same direction of the secondary waves. But the interference of all the secondary waves originating from the whole wave front is termed as diffraction. Hence the pattern obtained on the screen may be called an interference pattern or a diffraction pattern. The term interference may be used for those cases in which the resultant amplitude at a point is obtained by the superimposition of two or more beams. Diffraction can be defined as the phenomenon in which the resultant amplitude at any point on the screen is obtained by integrating the effect of infinitesimally small number of elements in to which the whole wave front can be divided. Thus, the resultant diffraction pattern obtained with a double slit can be taken as a combination of the effect of both interference and diffraction.

DIFFICULTIES WITH SIMPLE HUYGENS' PRINCIPLE

In Fig. 3.2 we have S as a source point, 145 an aperture, and SAC and SBD the projections within which we ordinarily expect light to be limited, if ray Optics is followed. But if we apply Huygens' principle, the secondary wavelets (as shown) should reach outside the limits AC and BD not marginally but to a considerable extent. But we know that diffraction effects in light occur only marginally. Ate we then to give up the Huygens' principle, or the wave theory, itself? Fresnel in 1815 showed by careful analysis that

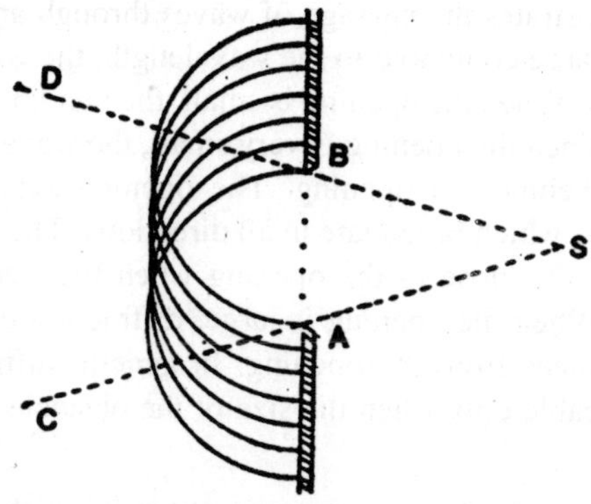

Fig. 3.2 : Explaining the difficulties with simple Huygens' principle.

(i) Huygens' wavelets cancel out for regions far outside AC and BD,

(ii) For regions well within AC and BD the Huygens' wavelets add up to give the same intensity as in the absence of the aperture, and

(iii) For regions close to the limits C and BD (both inside and outside) the wave theory predicts departures of intensity which match the experimental observations. In essence the theory involves summing up the contributions from different parts of the available wavefront, taking care of their phase relations.

One can argue that such 'summing up' is done to explain interference patterns also. The only difference is that in the diffraction case we have one continuous wavefront over which the effect is to be integrated, whereas in the interference case we have a discrete set of waves whose effect is to be summed up. So the difference is between summing up a series and integrating over some given limits. Some authors treat this distinction as trivial and use the word interference to cover all cases of super-positions of coherent waves. But we shall use the term diffraction as distinct from interference as a convenient usage.

DIFFRACTION

When waves encounter obstacles (or openings), they bend round the edges of the obstacles, if the dimensions of the obstacles are comparable to the wavelength of the waves. The bending of waves around the edges of an obstacle is called *diffraction.*

Fig. 3.3 illustrates the passage of waves through an opening. When the opening is large compared to the wavelength, the waves do not bend round the edges. When the opening is small, the bending round the edges is noticeable. When the opening is very small, the waves spread over the entire surface behind the opening. The opening acts an independent source of waves, which propagate in all directions. The diffraction effect is observable quite close to the opening when the size of the opening is very small. When the opening is large, diffraction effect is observed at greater distances from the opening. In general diffraction of waves becomes noticeable only when the size of the obstacle is comparable to a wavelength.

It is a matter of common experience that the path of light entering a dark room through a hole in the window illuminated by sunlight is straight. Similarly, if an opaque obstacle is placed in the path of light, a sharp shadow is cast on the screen, indicating thereby that light travels in straight lines. Rectilinear propagation of light can be easily explained on the basis of Newton's corpuscular theory. But it has been observed that when a beam of light passes through a small opening (a small circular hole or a narrow slit) it spreads to some extent into the region of the geometrical shadow also. If light energy is propagated in the form of waves, then similar to sound waves, one would expect bending of a beam of light round the edges of an opaque obstacle or illumination of the geometrical shadow.

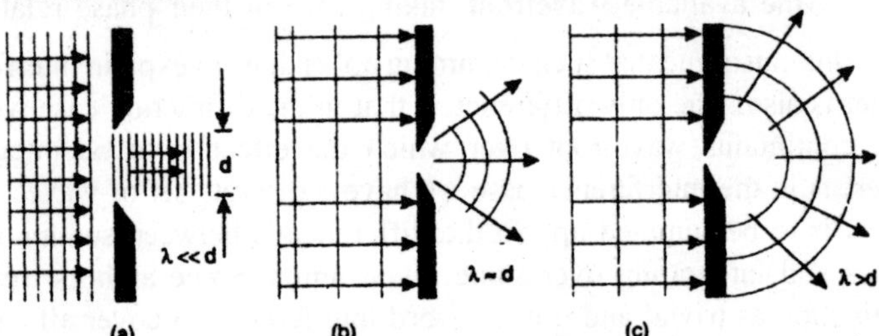

Fig. 3.3 : Diffraction—(a) A plane wave does not bend at the slit if the opening d ≈ λ. (b) Bending is perceptible when λ ≈ d (c). When λ > d, the bending takes place to such an extent that light can be perceived in a direction normal to the ray propagation suggesting that the opening acts as a point source.

However, diffraction phenomenon is not readily apparent in case of light waves. It becomes significant when the aperture size is of the order of one wavelength wide. Diffraction and interference are basically equivalent.

HUYGENS-FRESNEL THEORY

According to Huygen's wave theory of light, each progressive wave produces secondary waves, the envelope of which forms the secondary wave front. In Fig. 3.4(a), S is a source of monochromatic light and MN is a small aperture. XY is the screen placed in the path of light. AB is the illuminated portion of the screen and above A and below B is the region of the geometrical shadow. Considering MN as the primary wavefront, according to Huygen's construction, if secondary wave fronts are drawn, one would expect encroachment of light in the geometrical shadow. Thus, the shadows formed by small obstacles are not sharp. Bus bending of light round the edges of an obstacle or the encroachment of light within the geometrical, shadow is known as *diffraction*. Similarly, if an opaque obstacle MN is placed in the path of light Fig. 3.2,(b)], there should be illumination in the geometrical shadow region AB also. But the illuminated the geometrical shadow of an obstacle is not commonly observed because the light sources are not point sources and secondly the obstacles used are of very large size compared to the wavelength of light. If a shadow of an obstacle is cast by an extended source, say a frosted electric bulb, light from every point on the surface of the bulb forms its own diffraction pattern (bright and dark diffraction bands) and these overlap such that no single pattern can be identified. *The term diffraction is referred to such problems in which one considers the resultant effect produced by a limited portion of a wavefront.*

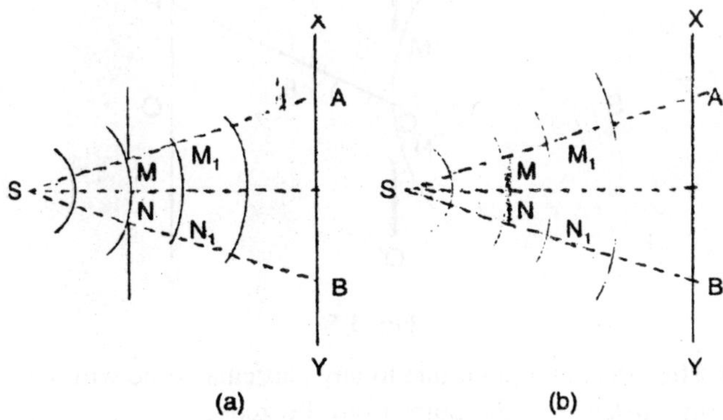

Fig. 3.4

Diffraction phenomena are part of our common experience. The luminous border that surrounds the profile of a mountain just before the sun rises behind it, the light streaks that one sees while looking at a strong source of light with half shut eyes and the coloured spectra (arranged in the form of a cross) that one sees while viewing a distant source of light through a fine piece of cloth are all examples of diffraction effects.

Augustine Jean Fresnel in 1815, combined in a striking manner Huygens' wavelets with the principle of interference and could satisfactorily explain the bending of light round obstacles and also the rectilinear propagation of light.

FRESNEL'S ASSUMPTIONS

According to Fresnel, the resultant effect at an external point due to a wavefront will depend on the factors discussed below:

In Fig. 3.5, S is a point source of monochromatic light and MN is a small aperture. XY is the screen and SO is perpendicular to XY. MCN is the incident spherical wavefront due to the point source S. To obtain the resultant effect at a point P on the screen, Fresnel assumed the following:

(1) A wave front can be divided into a large number of strips or *zones* called Presnel's zones of small area and the resultant effect at any point will depend on the combined effect of all the secondary waves emanating from the various zones;

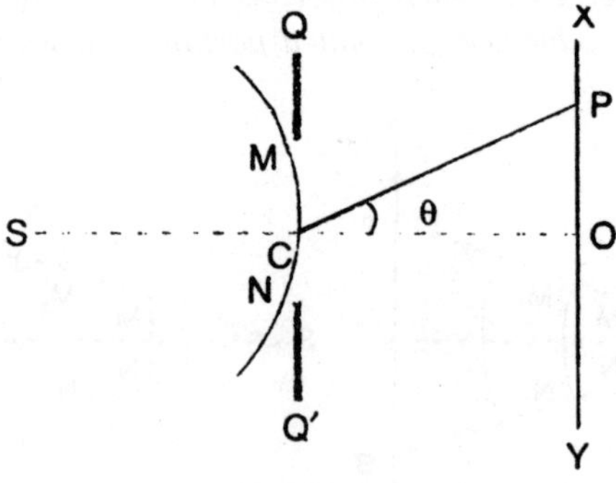

Fig. 3.5

(2) The effect at a point due to any particular zone will depend on the distance of the point from the zone;

(3) The effect at P will also depend on the obliquity of the point with reference to the zone under consideration, *e.g.* due to the part of the wavefront at C, the effect will be a maximum at O and decreases with increasing obliquity. It is a maximum in a direction radially outwards from C and it decreases in the opposite direction. The effect at a point due to the obliquity factor is proportional to $(1 + \cos \theta)$ where $\angle PCO = \theta$. Considering an elementary wavefront at C, the effect is maximum at O because $\theta = 0$ and $\cos \theta = 1$. Similarly, in a direction tangential to the primary wavefront at C (along CQ) the resultant effect is one half of that along CO because $\theta = 90°$ and $\cos 90° = 0$. In the direction CS, the resultant effect is zero since $\theta = 180°$ and $\cos 180° = -1$ and $1 + \cos 180° = 1 - 1 = 0$. This property of the secondary waves eliminates one of the difficulties experienced with the simpler form of Huygens principle viz., that if the secondary waves spread out in all directions from each point on the primary wavefront, they should give a wave travelling forward as well as backward. Now, as the amplitude at the rear of the wave is zero there will evidently be no back wave.

RECTILINEAR PROPAGATION OF LIGHT

ABCD is a plane wavefront perpendicular to the plane of the paper [Fig. 3.6 (a)] and P is an external point at a distance b perpendicular to ABCD. To find the resultant intensity at P due to the wavefront ABCD, Fresnel's method consists in dividing the wavefront into a number of *half period elements or zones* called Fresnel's zones and to find the effect of all the zones at the point P.

If spheres are constructed with P as centre and radii equal to $b + \lambda/2$, $b + 2\lambda/2$, $b + 3\lambda/2$ etc., they will cut out circular areas of radii OM_1, OM_2, OM_3, etc., on the wave front. These circular zones are called *half period zones* or *half period elements*. Each zone differs from its neighbour by a phase difference of π or path difference of $\lambda/1$.

Thus the secondary waves starting from the point O and M_1 and reaching P will have a phase difference of π or a path difference $\lambda/2$. A Fresnel half period zone with respect to an actual point P is a thin annular zone of the primary wavefront in which the secondary waves from any two corresponding points of neighbouring zones differ in path by $\lambda/2$.

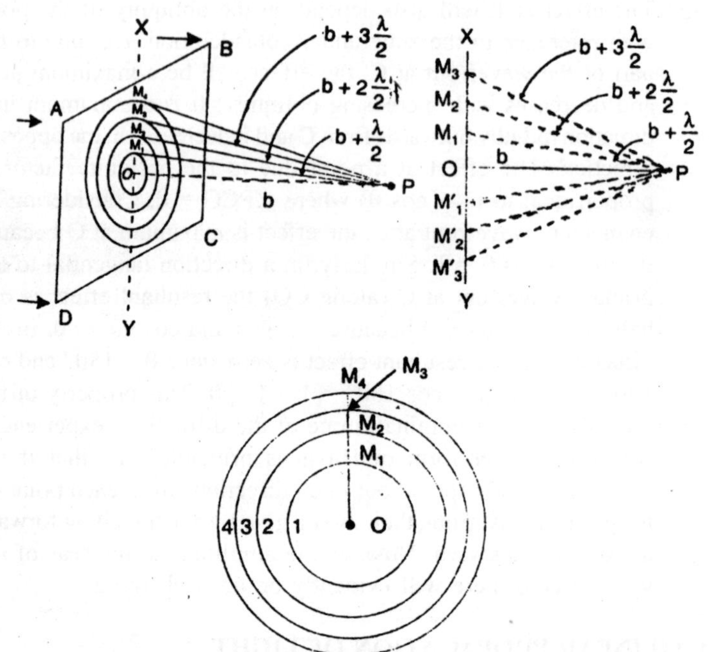

Fig. 3.6

In Fig. 3.6(b) O is the pole of the wavefront XY with reference to the external point P. OP is perpendicular to XY. In Fig. 3.6(c) 1, 2, 3 etc. are the half period zones constructed on the primary wavefront XY. OM_1 is the radius of the first zone. OM_2 is the radius of the second zone and so on. P is the point at which the resultant intensity has to be calculated.

$$OP = b,\ OM_1 = r_1,$$

$$OM_2, = r_2\ OM_3 = r_3,\text{ etc.}$$

$$\text{And}\ M_1P = b + \frac{\lambda}{2},\ M_2P = b + \frac{\lambda}{2},$$

$$M_3P = b + \frac{3\lambda}{2}\ \text{etc.}$$

The area of the first half period zone is

$$\pi\, OM_1^2 = \pi\left[M_1P^2 - OP^2\right] = \pi\left[\left(b + \frac{\lambda}{2}\right)^2 - b^2\right]$$

$$= \pi \left[b\lambda + \frac{\lambda^2}{4} \right] = \pi b\lambda \qquad ...(1)$$

As λ is small, λ^2 term is neglected.

The radius of the first half period zone is, $r_1 = OM_1 = \sqrt{b\lambda}$

The radius of the second half period zone is,

$$OM_2 = [M_2P^2 - OP^2]^{1/2}$$

$$= \left[(b + \lambda)^2 - b^2\right]^{1/2}$$

$$= \sqrt{2b\lambda}$$

The area of the second half period zone, $= \pi \left[OM_2^2 - OM_1^2\right]$

$$= \lambda[2b\lambda - b\lambda]$$

$$= \pi b\lambda \qquad ...(2)$$

Thus, the area of each half period zone is equal to π b λ. Also the radii of the 1st, 2nd, 3rd, etc. half period zones are $\sqrt{1b\lambda}$, $\sqrt{2b\lambda}$, $\sqrt{3b\lambda}$ etc. Therefore, the radii are proportional to the square roots of the natural numbers. However, it should be remembered that the areas of the zones are not constant but are dependent on– (i) λ, the wavelength of light and (ii) b, the distance of the point from the wavefront. The area of the zone increases with increase in the wavelength of light and with increase in the distance of the point P from the wavefront.

As discussed effect at a point will depend on (i) the distance of P from the wavefront, (ii) the area of the zone and (iii) the obliquity factor. Here, the area of each zone is the same.

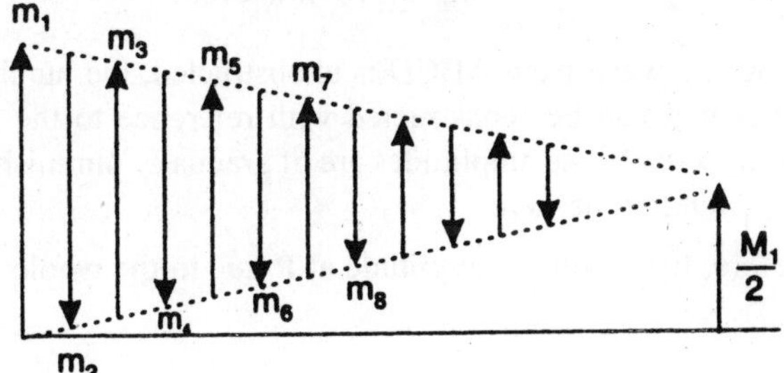

Fig. 3.7

The secondary waves reaching the point P are continuously out of phase and in phase with reference to the central or the first half period zone. Let m_1, m_2, m_3, etc. represent the amplitudes of vibration of the ether particles at P due to secondary waves from the 1st, 2nd, 3rd, etc. half period zones (Fig. 3.8). As we consider the zones outwards from O, the obliquity increases and hence the quantities m_1 m_2, m_3 etc. are of continuously decreasing order. Thus, m_1 is slightly greater than m_2; m_2 is slightly greater than m_3 and so on. Due to the phase difference of TC between any two consecutive zones, if the displacements of the ether particles due to odd numbered zones is in the positive direction, then due to the even numbered zones the displacement will be in the negative direction at the same instant. As the amplitudes are of gradually decreasing magnitude, the amplitude of vibration at P due to any zone can be approximately taken as the mean of the amplitudes due to the zones preceding and succeeding it.

e.g. $$m_2 = \frac{m_1 + m_3}{2}$$

The resultant amplitude at P at any instant is given by,

$$A = m_1 - m_2 + m_3 - m_4 \ldots\ldots + m_n \text{ if n is odd.}$$

(If n is even, the last quantity is $- m_n$).

$$\therefore \quad A = \frac{m_1}{2} + \left[\frac{m_1}{2} - m_2 + \frac{m_3}{2}\right] + \left[\frac{m_3}{2} - m_4 + \frac{m_5}{2}\right] + \ldots.$$

But $m_2 = \frac{m_2}{2} + \frac{m_3}{2}$ and $m_4 = \frac{m_3}{2} + \frac{m_4}{2}$

$$A = \frac{m_1}{2} + \frac{m_n}{2} \ldots\ldots\ldots \text{if n is odd.}$$

$$A = \frac{m_1}{2} + \frac{m_{n-1}}{2} - m_n \ldots\ldots\ldots \text{if n is even.}$$

If the whole wave front ABCD is unobstructed, the number of half period zones that can be constructed with reference to the point P is infinite *i.e.* $n \to \infty$. As the amplitudes are of gradually diminishing order, m_n and m_{n-1} tend to be zero.

Therefore, the resultant amplitude at P due to the whole wavefront

$$= A = \frac{m_1}{2}. \qquad \ldots(3)$$

The intensity at a point is proportional to the square of the amplitude.

$$\therefore \qquad I \propto \frac{m_1^2}{4} \qquad \text{...(4)}$$

Thus, the intensity at P is only one-fourth of that due to the first half period zone alone. Here, only half the area of the first half period zone is effective in producing the illumination at the point P. A small obstacle of the size of half the area of the first half period zone placed at O will screen the effect of the whole wavefront and the intensity at P due to the rest of the wavefront will be zero. While considering the rectilinear propagation of light, the size of the obstacle used is far greater than the area of the first half period zone and hence the bending effect of light round corners (diffraction effects) cannot be noticed. In the case of sound waves, the wavelengths are far greater than the wavelength of light, and hence the area of the first half period zone for a plane wavefront of sound is very large. If the effect of sound at a point beyond an obstacle is to be shadowed, an obstacle of very large size has to be used to get no sound effect. If the size of the obstacles placed in the path of light is comparable to the wavelength of light, then it is possible to observe illumination in the region of the geometrical shadow also. Thus, rectilinear propagation of light is only approximately true.

ECHELON GRATING

An echelon transmission grating consists of a number of optically worked glass plates arranged in the form of steps, as shown in Fig. 3.8. All the plates are cut from a single optically worked glass plate. Each plate overlaps on the next by the same distance *i.e.*, the step- width is the same throughout and is of the order of 1mm. A parallel beam of monochromatic light incident normally is diffracted through a small angle.

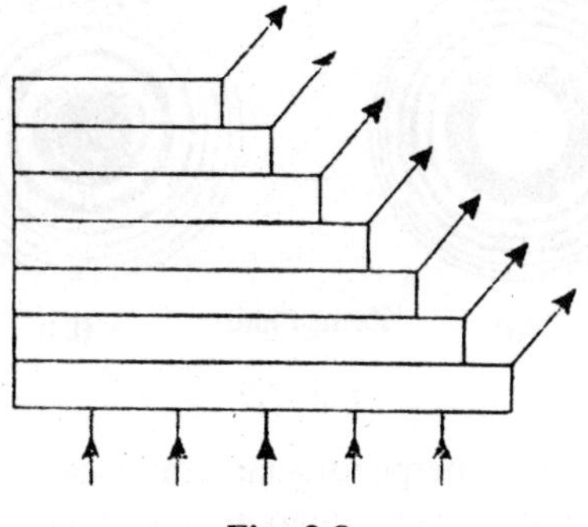

Fig. 3.8

One can observe the fifth or sixth order spectrum with a concave grating and only the second or third order spectrum with a plane diffraction grating. The resolving power of a ruled grating is dependent on the number of lines on the grating surface and the order of the spectrum. It is difficult and very tedious to draw a large number of equidistant parallel lines on a grating surface. With an echelon grating, designed by Michelson, one can observe the spectrum of a very high order and hence the resolving power of the echelon grating is very high. But, with the increase in the order of the spectrum, the intensity of the spectral lines decreases considerably. Also, the angular spacing $d\theta$ is very small and hence there will be many over lapping orders. An echelon grating, therefore, is not suitable to study the spectrum as such but essentially helps in detecting the true monocromatism of a beam of light.

ZONE PLATE

A zone plate is a specially constructed screen such that light is obstructed from every alternate zone. It can be designed so as to cut off light due to the even numbered zones or that due to the odd numbered zones. The correctness of Fresnel's method in dividing a wavefront into half period zones can be verified with its help.

To construct a zone plate, concentric circles are drawn on white paper such that the radii are proportional to the square roots of the natural numbers the radii are proportional to the square roots of the natural numbers). The odd numbered zones (*i.e.* 1st, 3rd, 5th, etc.) are covered with black ink and a reduced photograph is taken. The drawing appears as shown in Fig. 3.9(b). The negative of the photograph will be as shown in Fig. 3.9(a). In the developed negative, the odd zones are transparent to incident light and the even zones will cut off light.

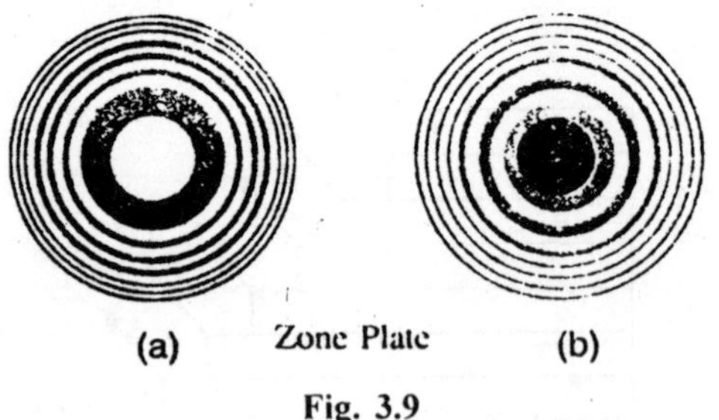

Zone Plate

Fig. 3.9

If such a plate is held perpendicular to an incident beam of light and a screen is moved on the other side to get the image, it will be observed that maximum brightness is possible at some position of the screen say b cm from the zone plate (Fig. 3.10) XO is the upper half of the incident plane wavefront. P is the point at which the light intensity is to be considered. The distance of the point P from the wavefront is b. OM_1(= r_1), OM_2(= r_2) etc. are the radii of the zones.

$r_1 = \sqrt{b\lambda}$ and $r_2 \sqrt{2b\lambda}$ where λ is the wavelength of light.

$$r_n = \sqrt{nb\lambda} \text{ or } b = \frac{r_n^2}{m\lambda} \quad ...(5)$$

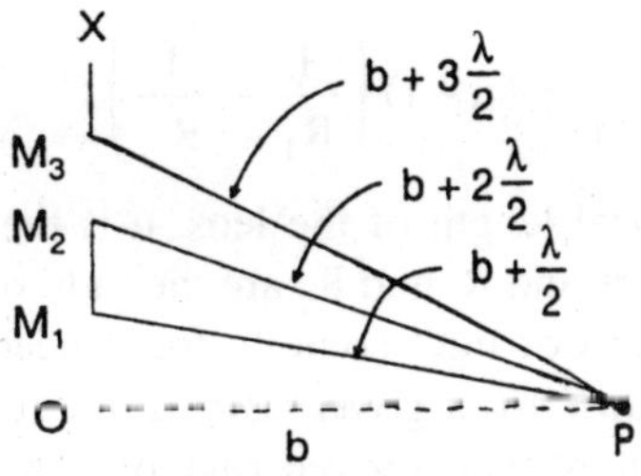

Fig. 3.10

If the source is at a large distance from the zone plate, a bright spot will be obtained at P. As the distance of the source is large, the incident wavefront can be taken as a plane one with respect to the small area of the zone plate. The even numbered zones cut off the light and hence resultant amplitude at P = A = $m_1 + m_2 + m_3$ +..... etc. In this case the focal length of the zone plater f_n is given by

$$f_n = b = \frac{r_n^2}{n\lambda} \quad ...(6)$$

Thus, a zone plate has different foci for different wavelengths. The- radius of the nth zone increases with increasing value of λ. *It is very interesting to note that as the even numbered zones are opaque, the intensity at P is much greater than that when the whole wavefront is exposed to the point P.*

In the first case the resultant amplitude is given by

$$A = m_1 + m_3 + m_5 + ... m_n.... \quad \text{(n is odd)}$$

When the whole wavefront is unobstructed, the amplitude is given by

$$A = m_1 - m_2 + m_3 - m_4 ...... + m_n$$

$$= \frac{m_1}{2} \text{ (if n is very large and n is odd).}$$

If a parallel beam of white light is incident on the zone plate, different colours come to focus at different points along the line OP. Thus, the function of a *zone plate is similar to that of a convex* (converging) lens and a formula connecting the distance of the object and image points can be obtained for a zone plate also.

Difference Between a Zone Plate and a Convex Lens

For a given wavelength of light, a convex lens has only one focal length given by

$$\frac{1}{f} = (\mu - 1)\left(\frac{1}{R_1} - \frac{1}{R_2}\right)$$

where f is the focal length of the lens, u is the refractive index of the material of the lens and R, and R_2 are the radii of curvature. In a convex lens, the violet rays of light come to focus nearer the lens than the red rays of light because for a given material the refractive index for violet rays of light is more than for red rays of light.

In the case of a zone plate, there are a number of foci between the point O and P. Each focus corresponds to the position where, with reference to P an odd number of half period elements can be constructed on each zone. As the screen is moved nearer the zone plate, the area of the half period elements decreases and more half period elements can be present on each zone. If P_m is the position on the image when $(2_m - 1)$ half period elements can be present on each zone, f_m the focal length of the zone plate is given by

$$f_m = \frac{r_n^2}{(2m - 1)\, n\lambda} \qquad ...(7)$$

Putting m = 1, 2, 3.., etc., the different positions of the screen for a bright image can be obtained. In equation (7), r is the radius of the nth zone of the wavefront, λ is the wavelength of light and (2m – 1) is the number of odd half period elements present on each zone. For example, if the position of the screen is such that with reference to the point P, three half period elements can be constructed on each zone, then the focal length of the zone plate f_3 is given by

$$f_3 = \frac{r_n^2}{5n\lambda}$$

With the decrease in the focal length of the zone plate, the brightness of the image decreases. Let the first zone contain only one period element.

Then, the amplitude at P due to this zone is m_1. If the first zone contains three half period elements for a particular position of the screen, then the amplitude at P due to the first zone

$$= m_1 - m_2 + m_3$$

$$= \frac{m_1}{2} + \left[\frac{m_1}{2} - m_2 + \frac{m_3}{2}\right] + \frac{m_3}{2} = \frac{m_1}{2} + \frac{m_2}{2}$$

But, $\frac{m_1 + m_3}{2}$ is less than m_1 because $m_1 > m_3$.

Further, in a zone plate (for the same number of odd half period elements contained in each zone) the focal length for violet light is more than for red light, which is reverse in the case of a convex lens.

Comparison Between a Zone Plate and a Convex Lens

1. Both the zone plate and convex lens form a real image of the object and the equations connecting the conjugate distances are similar.
2. The focal lengths of both depend on the wavelength, λ and hence suffer from chromatic aberration. The chromatic aberration in a zone plate is much more severe than in a convex lens.
3. A zone plate acts simultaneously as a convex lens and as a concave lens. In addition to a real image, a virtual image is also formed simultaneously. A convex lens forms only a real image.
4. In case of zone plate the image is formed by the diffraction phenomenon. In case of a convex lens the image is formed due to refraction of light.
5. The zone plate has got multiple foci on either side of the plate. Hence, the intensity of the image formed will be much less. Convex lens has only one focus. As all the light is focused at one point, the intensity of the image will be more.
6. In a zone plate, waves reaching the image point through any two alternate zones differ in path by λ, and in phase by 2π. In case of a convex lens all the rays reaching the image point have zero path or phase difference.
7. A zone plate can be used over a wide range of wavelengths from microwaves to x-rays. Glass lens cannot be used beyond the visible region.

CONCAVE REFLECTION GRATING

The wavelength of a spectral line can be determined accurately with a plane transmission grating, knowing the grating constant (a + b), the diffraction angle θ and the order n. From the knowledge of the wavelength of a single line, the wavelengths of the other lines can be obtained by comparison. Use of a plane transmission grating requires two lenses, the collimating lens and the telescope objective. The collimating lens gives a parallel beam of light and the telescope objective focuses the diffracted beam. The use of two lenses, if they are not perfectly achromatic, makes the spectrum more complex due to the chromatic aberration present in the lenses. Rowland developed the *concave reflection grating*, the use of which dispenses the use of the lenses. The rulings are made on a concave reflecting surface instead of a plane surface. The concave mirror is highly polished metal surface and it will diffract the incident beam and also focuses it at the same time. In a concave reflection grating, the effect of chromatic aberration is completely eliminated and it can be conveniently used in those regions of the spectrum for which the glass lenses are not transparent.

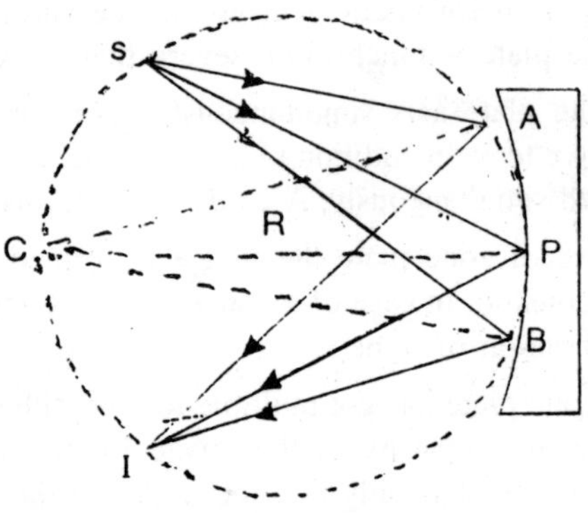

Fig. 3.11

In Fig. 3.11, APB is the surface of a concave reflection grating in which the rulings are perpendicular to the plane of the paper. C is the centre of curvature of the surface, that is CP = R. The dotted circle is the Rowland circle, which has a diameter R. The circle touches the grating surface at P. If a source of light S is placed at any point on the circumference of the Rowland circle PSCI, the dispersed spectral images of the slit are obtained at points such as I on the circumference of the same circle. By keeping the source of light at C, the spectra can be observed at other points on the circumference of the circle.

PASCHEN MOUNTING

The common form of mounting used for a concave reflection grating is shown in Fig. 3.12. It is called the *Paschen mounting.*

In this mounting, the slit S is set on the circumference of the Rowland circle as shown. The slit is perpendicular to the plane of the paper. GG' is the concave reflection grating and OC' is the diameter of the Rowland circle. The spectra of different orders are imaged on the circumference of the circle. In the figure, C is the central image; A_1B_1 is the first order spectrum. A_2B_2 is second order spectrum and A_3B_3 is the third order spectrum. With this mounting, several orders of the spectrum can be photographed simultaneously. The photographic plates are held in a frame which can give the plates the proper curvature coincident with the Rowland circle. For any particular order, the dispersion is minimum when $\theta = 0$. The disadvantage in this mounting is that the spacing of the spectral lines in the different regions of the spectrum is not proportional to the difference in wavelength between the lines.

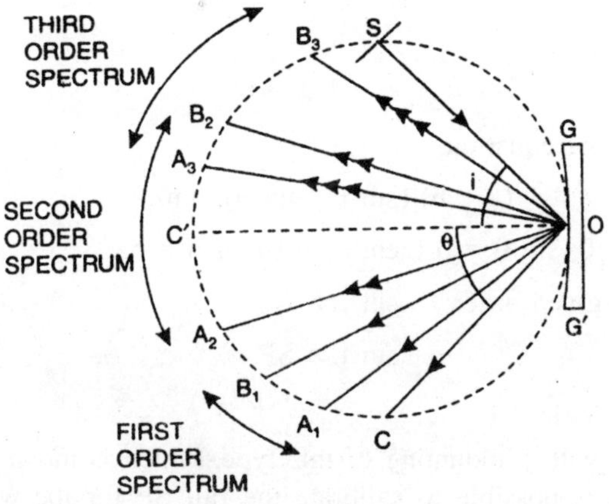

Fig. 3.12

ROWLAND MOUNTING

The principle of Rowland mounting is illustrated in Fig. 7.17. G is the concave grating and P is the plate holder. The grating and plate holder are mounted at the ends of a beam of length R equal to the radius of curvature of the grating surface. This beam GP can slide along two rails SX and SY. G, P and G', P', represent two positions of the beam. The slit S is set at the point of intersection of the rails SX and SY. With an arrangement of this type, the region of the spectrum imaged at P can be altered by sliding the beam. Sliding the beam alters the angle of incidence i. The spectrum obtained with this arrangement is nearly normal because the angle θ is nearly zero. For any position of the beam the spectrum is imaged at P.

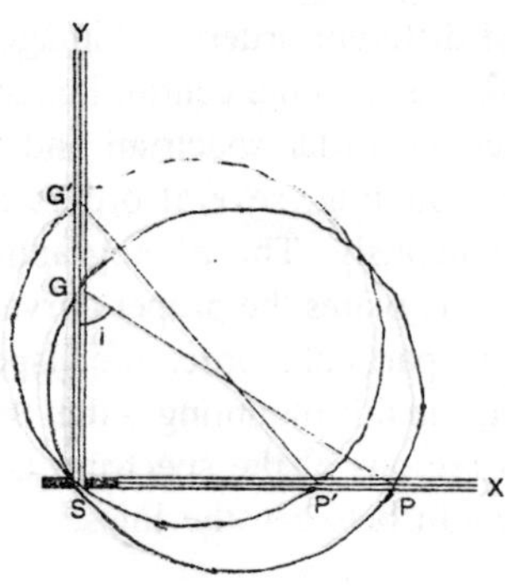

Fig. 3.13

From the equation,

$$(a + b)(\sin i - \sin \theta) = n\lambda$$

if $\theta = 0$; $\sin \theta = 0$ then, $(a + b) \sin i = n\lambda$

For a given order $\quad \sin i \propto \lambda$

But $\quad \sin i \propto SP$

$\therefore$

$$SP \propto \lambda \qquad \text{...(8)}$$

Thus, with a mounting of this type, which is mostly of historical interest, it is possible to calibrate the rail SP for the wavelengths of spectral lines.

EAGLE MOUNTING

The Rowland and Paschen mountings have largely been replaced by the Eagle mounting illustrated in Fig. 3.14. In this mounting, that portion of the spectrum which is diffracted back at an angle almost equal the angle of incidence, is focused on the plate P. To study the different regions of the spectrum the grating is turned about an axis perpendicular to the plane of the paper. Correspondingly, the plate holder P, which is hinged on one side of S is turned such that P and S lie on the Rowland circle. For studying spectra in the ultra violet region, the Eagle mounting is commonly used in vacuum spectrographs.

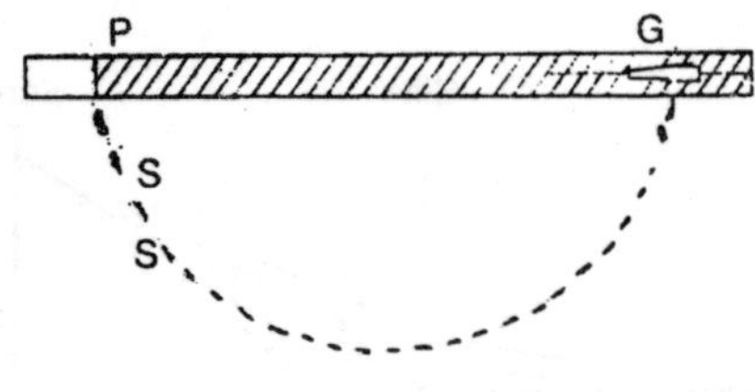

Fig. 3.14

LITTROW MOUNTING

Littrow mounting is illustrated in Fig. 3.15. Large plane reflection gratings are mounted this way. G is a plane reflection grating, P is a photographic plate, S is a slit and L is a large achromatic lens. The principle of Littrow mounting is similar to that of Eagle mounting.

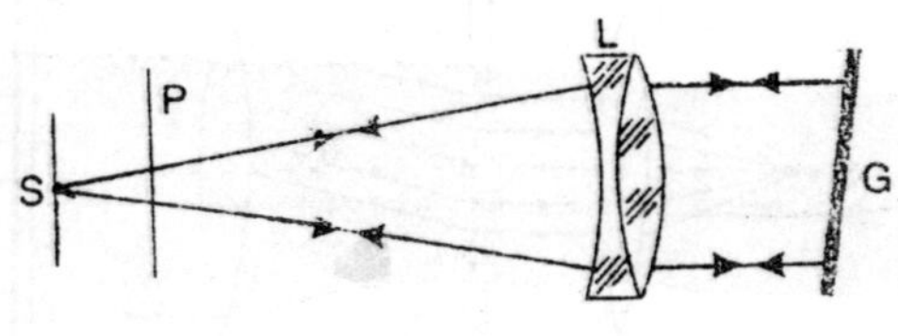

Fig. 3.15

The achromatic lens L serves two functions. It renders the incident light from the slit parallel as well as it focuses the diffracted beam on P. Thus, it serves both as collimating lens and a telescope objective.

FRESNEL AND FRAUNHOFFER TYPES OF DIFFRACTION

The diffraction phenomena are broadly classified into two types: Fresnel diffraction and Fraunhoffer diffraction.

1. *Fresnel Diffraction :* In this type of diffraction, the source of light and the screen are effectively at finite distances from the obstacle (Fig. 3.16a). Observation of Fresnel diffraction phenomenon does not require any lenses. The incident wave

front is not planar. As a result, the phase of secondary wavelets is not the same at all points in the plane of the obstacle. The resultant amplitude at any point of the screen is obtained by the mutual interference of secondary wavelets from different elements of unblocked portions of wave front. It is experimentally simple but the analysis proves to be very complex.

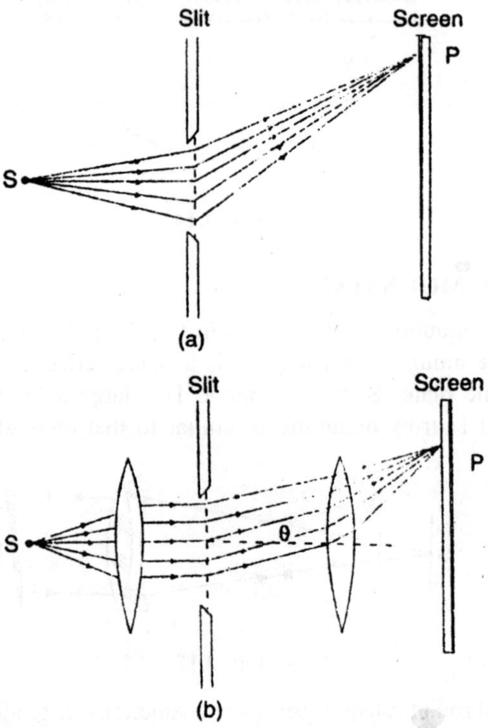

Fig. 3.16 : Conditions for Fresnel diffraction and Fraunhoffer diffraction.

2. *Fraunhoffer Diffraction :* In this type of diffraction, the source of light and the screen are effectively at infinite distances from the obstacle. Fraunhoffer diffraction pattern can be easily observed in practice. The conditions required for Fraunhoffer diffraction are achieved using two convex lenses, one to make the light from the source parallel and the other to focus the light after diffraction on to the screen (Fig. 3.16b). The diffraction is thus produced by the interference between parallel rays. The incident wave

front as such is plane and the secondary wavelets, which originate from the unblocked portions of the wave front, are in the same phase at every point in the plane of the obstacle. This problem is simple to handle mathematically because the rays are parallel. The incoming light is rendered parallel with a lens and diffracted beam is focused on the screen with another lens.

Fresnel class of diffraction phenomenon is treated in this chapter.

DIFFRACTION AT A CIRCULAR APERTURE

Let AB be a small aperture (say a pin hole) and S is a point source of monochromatic light. XY is a screen perpendicular to the plane of the paper and P is a point on the screen. SP is perpendicular to the screen O is the center of the aperture and r is the radius of the aperture.

Let the distance of a source from the aperture be a (SO = a) and the distance of the screen from the aperture be b (OP = b). QOQ_1 is the incident spherical wavefront and with reference to the point P, O is the pole of the wavefront (Fig. 3.17). To consider the intensity at P, half period zones can be constructed with P as center and radii $b + \frac{\lambda}{2}$, $b + \frac{2\lambda}{2}$ etc. on the exposed wavefront AOB. Depending on the distance of P from the aperture (*i.e.* the distance b) the number of half period zones that can be constructed may be odd or even. If the distance a is such that only one half period zone can be constructed, then the intensity at P will be proportional to m_1^2 (where m_1 is the amplitude due to the first zone at P). On the other hand, if the whole of the wavefront is exposed to the point P, the resultant amplitude is $\frac{m_1}{2}$ or the intensity at P will be proportional to $\frac{m_1^2}{4}$. The position of the screen can be altered so as to construct 2, 3 or more half period zones for the same area of the aperture. If only two zones are exposed, the resultant amplitude at P = $m_1 - m_2$ (minimum) and if 3 zones are exposed, the amplitude = $m_1 - m_2 + m_3$, (maximum) and so on. Thus by continuously altering the value of b, the point P becomes alternately bright and dark depending on whether odd or even number of zones are exposed by the aperture.

Now let us consider a point P' on the screen XY (Fig. 3.18). Let S to P' be Joined. The line SP' meets the wavefront at O'.O' is the pole of the wavefront with reference to the point P'. Now let us construct half period zones with the point O' as the pole of the wavefront. The upper half of the wavefront is cut off by the obstacle.

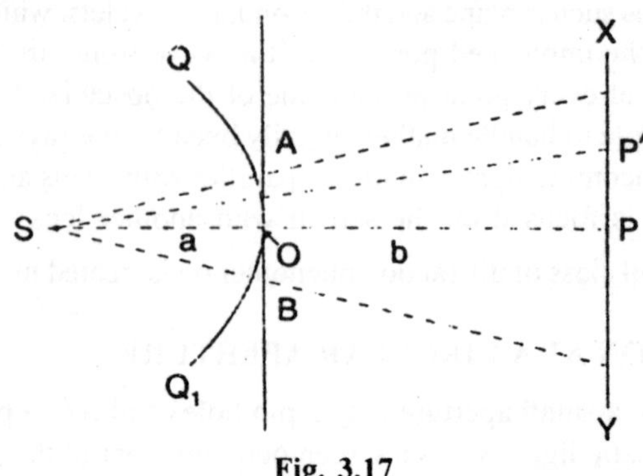

Fig. 3.17

If the first two zones are cut off by the obstacle between the points O' and A and if only the 3rd, 4th and 5th zones are exposed by the aperture AOB, then intensity at P' will be maximum. Thus if odd number of half period zones are exposed, point P' will be of maximum intensity and if even number of zones are exposed, point P' will be of minimum intensity. As the distance of P' from P increases the intensity of maxima and minima gradually decreases. It is because with the point P' far removed from P, the most effective central half period zones are cut off by the obstacle between the points O' and A. With the outer zones the obliquity increases with reference to the point P' and hence the intensity of maxima and minima also will be less. If the point P' happens to be of maximum intensity, then all the points lying on a circle of radius PP' on the screen also will be of maximum intensity.

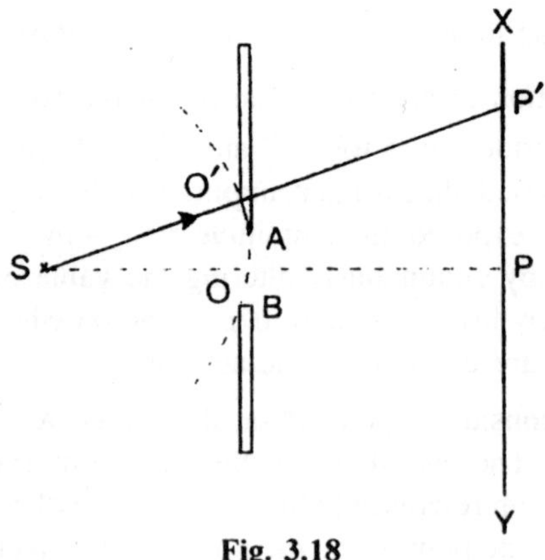

Fig. 3.18

Thus with a circular aperture, the diffraction pattern will be concentric bright and dark rings with the centre P bright or dark depending on the distance b. The width of the rings continuously decreases.

Mathematical Treatment of Diffraction at a Circular Aperture

In Fig. 3.19 S is a point source of a monochromatic light, AB is the circular aperture and P is point on the screen O is the center of the circular aperture. The line SOP is perpendicular to the circular aperture AB and the screen at P. The screen is perpendicular to the plane of the paper.

Let δ be the path difference for the wave reaching P along the paths SAP and SOP.

$$SO = a;\ OP = b;\ OA = r$$

$$\delta = SA + AP - SOP$$

$$= \left(a^2 + r^2\right)^{1/2} + \left(b^2 + r^2\right)^{1/2} - (a + b)$$

$$= a\left(1 + \frac{r^2}{a^2}\right)^{1/2} + b\left(1 + \frac{r^2}{b^2}\right)^{1/2} - (a + b)$$

$$= a\left(1 + \frac{r^2}{2a^2}\right) + b\left(1 + \frac{r^2}{2b^2}\right) - (a + b) = \frac{r^2}{2}\left(\frac{1}{a} + \frac{1}{b}\right) \quad ...(9)$$

$$\therefore \quad \frac{1}{a} + \frac{1}{b} = \frac{2\delta}{r^2}$$

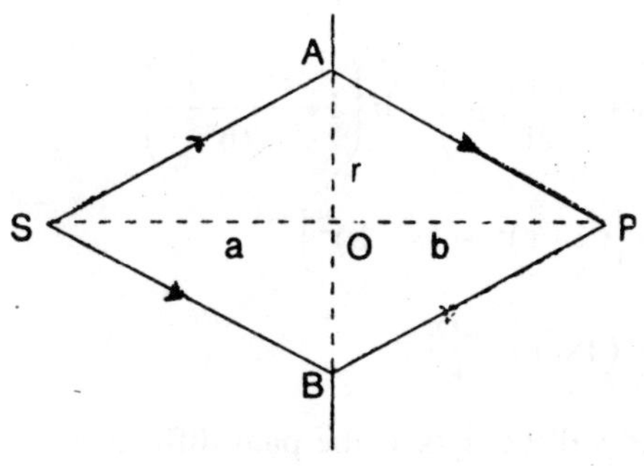

Fig. 3.19

If the position of the screen is such that n full number of half period zones can be constructed on the aperture, then the path difference,

$$\delta = \frac{n\lambda}{2} \text{ or } 2\delta = n\lambda,$$

Substituting the value of 28 in (8), we get,

$$\frac{1}{a} + \frac{1}{b} = \frac{n\lambda}{r^2} \qquad ...(10)$$

The point P will be of maximum or minimum intensity depending on whether n is odd or even. If the source is at infinite distance (for an incident plane wave front), then $\alpha = \infty$ and

$$\frac{1}{b} = \frac{1}{f} = \frac{n\lambda}{r^2} \qquad ...(11)$$

If n is odd, P will be a bright point. The idea of focus at P does not mean that it is always a bright point.

Intensity at a Point Away from the Centre

In Fig. 3.20 AB is a circular aperture and P and P' are two points on the screen. PP' = x and OP = b. OP is perpendicular to the screen.

Let r be the radius of the aperture. The path difference between the secondary waves from A and B and reaching P' can be given by,

$$\delta = BP' - AP'$$

$$= \sqrt{b^2 + (x+r)^2} - \sqrt{b^2 + (x-r)^2}$$

$$= b\left(1 + \frac{(x+r)^2}{2b^2}\right) - b\left(1 + \frac{(x-r)^2}{2b^2}\right)$$

$$= \left(b + \frac{(x+r)^2}{2b^2}\right) - b\left(1 + \frac{(x-r)^2}{2b^2}\right)$$

$$= \frac{1}{2b}\left[(x+r)^2 - (x-b)^2\right]$$

$$\therefore \quad \delta = \frac{1}{2b}(4xr) = \frac{2rx}{b} \qquad ...(12)$$

The point P' will be dark if the path difference,

$\delta = 2n\frac{\lambda}{2}$ (2n means even number of zones).

$$2n\frac{\lambda}{2} = \frac{2rx_n}{b}$$

or $$x_n = \frac{nb\lambda}{2r} \quad ...(13)$$

where x_n gives the radius of nth dark ring.

Similarly, if $\delta = \frac{(2n+1)\lambda}{2}$,

then $$\frac{(2n+1)\lambda}{2} = \frac{2rx_n}{b}$$

or $$x_n = \frac{(2n+1)b\lambda}{4r} \quad ...(14)$$

where x_n gives the radius of the nth bright ring.

The objective of a telescope consists of an achromatic convex lens and a circular aperture is fixed in front of the lens. Let the diameter of the aperture be D (= 2r). While viewing distant objects, the incident wave front is plane and the diffraction pattern consists of a bright centre surrounded by dark and bright rings of gradually decreasing intensity. The radii of the dark rings are given by

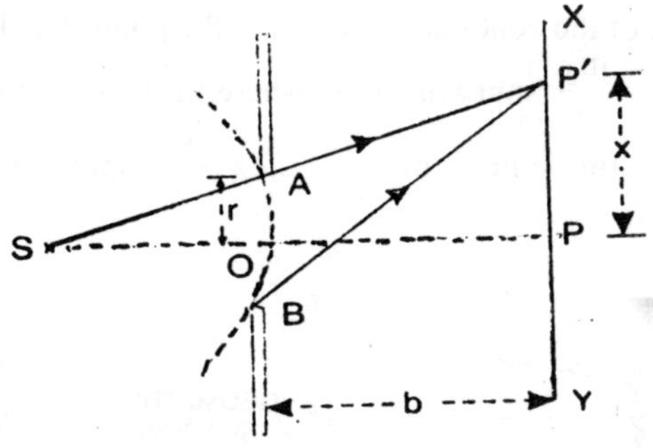

Fig. 3.20

$$x_n = \frac{nb\lambda}{2r} = \frac{nb\lambda}{D} \quad ...(15)$$

The radius of the first dark ring is, $x_1 = \frac{b\lambda}{D}$

For an incident plane wavefront, b = f the focal length of the objective.

$$\therefore \quad x_1 = \frac{f^{\lambda}}{D}$$

The value of x_1 measures the distance of the first secondary minimum from the central bright maximum. However, according to Airy's theory, the radius of the first dark ring is given by

$$x_1 = \frac{1.22 f\lambda}{D} \qquad ...(16)$$

It is interesting to note that the size of the central image depends on λ, the wavelength of light, f, the focal length of the lens and D, diameter of the lens aperture.

DIFFRACTION AT AN OPAQUE CIRCULAR DISC

S is a point source of monochromatic light. CD is an opaque disc and MN is the screen. P is a point on the screen such that SAP is perpendicular to the screen. The screen is perpendicular to the plane of the paper. XY is the incident spherical wave front. EF is the geometrical shadow. With reference to the point P, the wave front can be divided into half period zones taking the centre of the disc (A) as the pole (Fig. 3.21). If one half period zone can be constructed on the surface of the disc the rest of the zones are exposed to the point P and the resultant amplitude at P $= \frac{m_2}{2}$ approximately, where m_2 is the amplitude due to the second zone $[m_2 - m_3 + m_4 - .... = \frac{m_2}{2}$ approximately].

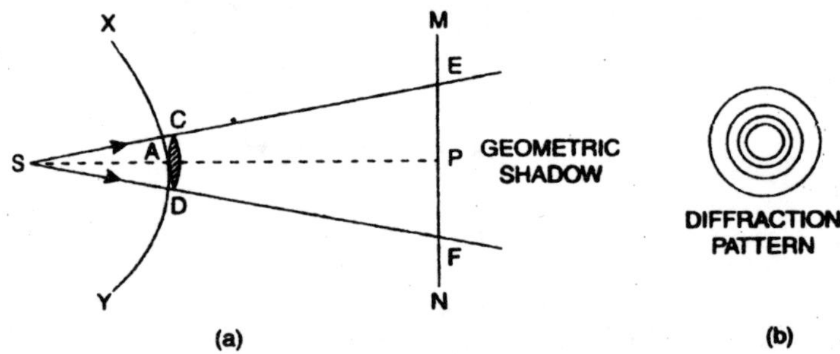

Fig. 3.21

Similarly, if two half period zones can be constructed on the surface of the disc, the resultant amplitude at P due to the exposed zones will be $= \frac{m_3}{2}$ and so on. Thus, the point P will always be bright but the intensity at P decreases with increase in the diameter of the disc. That is, with a large diameter of the disc, the most effective central zones will be cut off by the disc and the exposed outer zones are more oblique with reference to the point P. Thus, (at P) the center of the geometrical shadow will be bright as if the disc were absent. The diffraction pattern consists of a central bright spot surrounded by alternate bright and dark rings, as shown in Fig. 3.21(b).

DIFFRACTION PATTERN DUE TO A STRAIGHT EDGE

Let S be narrow slit illuminated by a source of monochromatic light of wavelength, λ. The length of the slit is perpendicular to the plane of the paper. AD is the straight edge and the length of the edge is parallel to the length of slit (Fig. 3.22). XY is the incident cylindrical wavefront. P is a point on the screen and SAP is perpendicular to the screen. The screen is perpendicular to the plane of the paper.

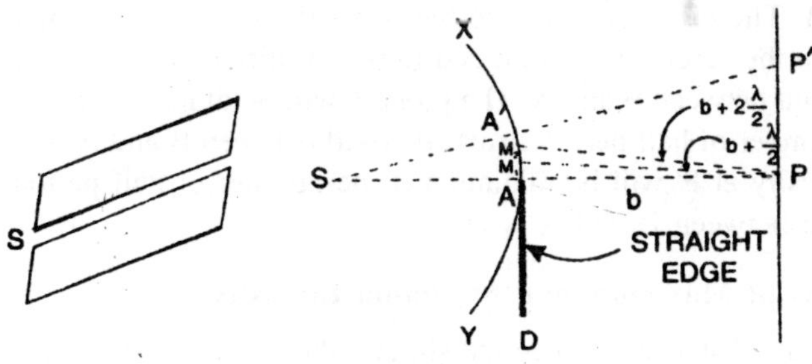

Fig. 3.22

Below the point P is the geometrical shadow and above P is the illuminated portion. Let the distance AP be b. With reference to the point P, the wave front can be divided into a number of half period strips, as shown in Fig. 3.23.

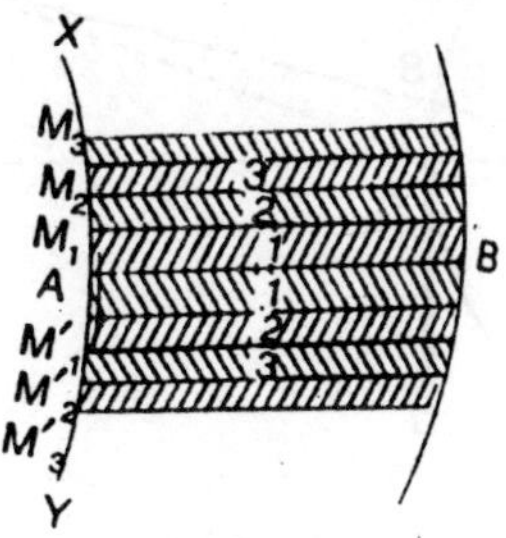

Fig. 3.23

XY is the wave front, A is the pole of the wave front and AM_1, M_1M_2, M_2M_3 etc. measure of the thickness of the 1st, 2nd, 3rd etc. half period strips. With the increase in the order of the strip, the area of the strip decreases (Fig. 3.23).

In Fig. 3.22, AP = b,

$$PM_1 = b + \frac{\lambda}{2}$$

and $PM_2 = b + \frac{2\lambda}{2}$ etc.

Let P' be a point on the screen in the illuminated portion (Fig. 3.24). To calculate the resultant effect at P' due to the wave front XY, let us join S to P'. This line meets the wave front at B. B is the pole of the wave front with reference to the point P' and the intensity at P' will depend mainly on the number of half period strips enclosed between the points A and B. The effect at P' due to the wave front above B is same at all points on the screen whereas it is different at different points due to the wave front between B and A. The point P' will be of maximum intensity, if the number of half period strips enclosed between B and A is odd and the intensity at P' will be minimum if the number of half period strips enclosed between B and A is even.

Positions of Maximum and Minimum Intensity

Let the distance between the slit and the straight edge be a and the distance between the straight edge and the screen be b (Fig. 3.24). Let PP' be x.

The path difference, δ = AP' – BP'

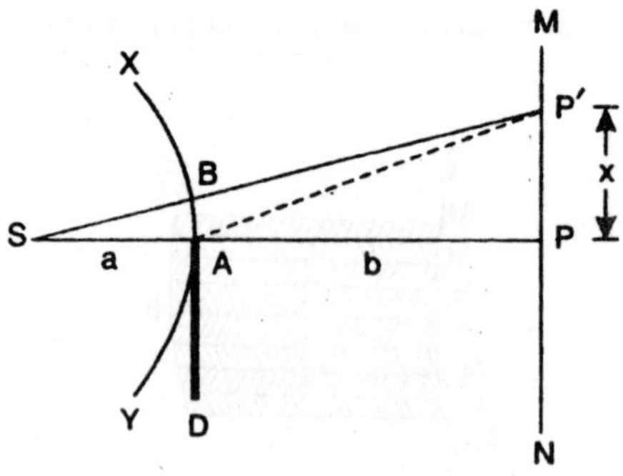

Fig. 3.24

$$= (b^2 + x^2)^{1/2} = [SP' + SB$$

$$= (a^2 + b^{2/12} - \left(\sqrt{(a+b)^2 + x^2 - a}\right)$$

$$= b\left[1 + \frac{x^2}{2b^2}\right] - (a+b)\left[1 + \frac{x^2}{2(a+b)^2}\right] + a$$

$$= \frac{x^2}{2}\left(\frac{1}{b} - \frac{1}{a+b}\right) = \frac{x^2}{2}\left(\frac{a+b-b}{b(a+b)}\right)$$

$$\therefore \quad \delta = \frac{x^2}{2} \cdot \frac{a}{b(a+b)}$$

The point P' will be of maximum intensity if $\delta = (2n+1)\frac{\lambda}{2}$

$$\therefore \quad (2n+1)\frac{\lambda}{2} = \frac{ax_x^2}{2b(a+b)}$$

$$x_n = \sqrt{\frac{(2n+1)\,(a+b)\,b\lambda}{2}} \qquad \text{...(17)}$$

where x_n is the distance of the nth bright band from P.

Similarly, P' will be of minimum intensity if $d = 2n\frac{\lambda}{2}$

$$\therefore \qquad 2n\,\frac{\lambda}{2} = \frac{ax_n^2}{2b(a+b)}$$

or

$$x_n = \frac{\sqrt{2n(a+b)b\lambda}}{a}$$

where x_n is the distance of the nth dark band from P. Thus, diffraction bands of varying intensity (roughly corresponding to maxima and minima) are observed above the geometrical shadow *i.e.*, above P and the bands disappear and uniform illumination occurs if P' is far away from P.

Intensity at a Point Inside the Geometrical Shadow (Straight Edge)

If P' is a point below P (Fig. 3.25) and B is the new pole of the wave front with reference to the point P', then the half period strips below B are cut off by the obstacle and only the uncovered half period strips above B will be effective in producing the illumination at P'. As P' moves farther from P, more number of half period strips above B is also cut off and the intensity gradually falls. Thus within the geometrical shadow, the intensity gradually falls off depending on the position of P' with respect to P.

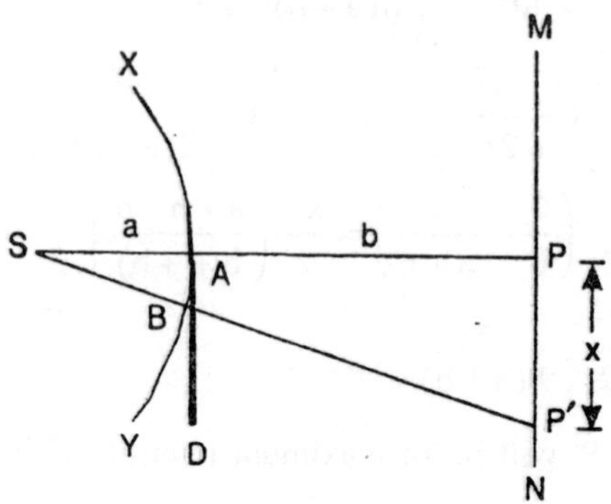

Fig. 3.25

The intensity distribution on the screen due to a straight edge is shown in Fig. 3.26. S is the source, AD is the straight edge and MN is the screen. In the illuminated portion PM, alternate bright and dark bands of gradually diminishing intensity will be observed and the intensity falls off gradually in the region of the geometrical shadow.

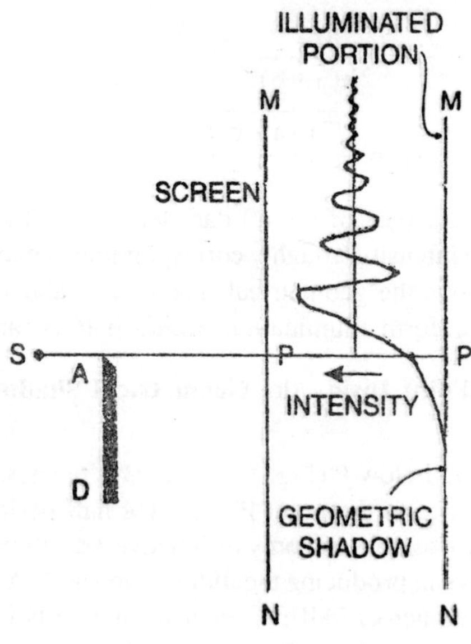

Fig. 3.26

Thus according to the wave theory, the shadows cast by obstacles in the path of light are not sharp and hence rectilinear propagation of light is only approximately true. In general, there is gradual fading of intensity in the region of the geometrical shadow and with monochromatic light bright and dark bands (diffraction bands) are observed in the illuminated portion of the screen. However, with white light coloured bands will be observed and the bands of shorter wavelength are nearer the point P.

FRESNEL AND FRAUNHOFFER CLASSES OF DIFFRACTION

In Fig. 3.27 (top), A' B' and C' D' are projections of the apertures AB and CD which limit the wavefront. The gross structure of A' B' (for instance) matches AB but has nothing to do with the structure of the source S. The modifications in detailed structure of A' B' are called Fresnel diffraction, which we studied.

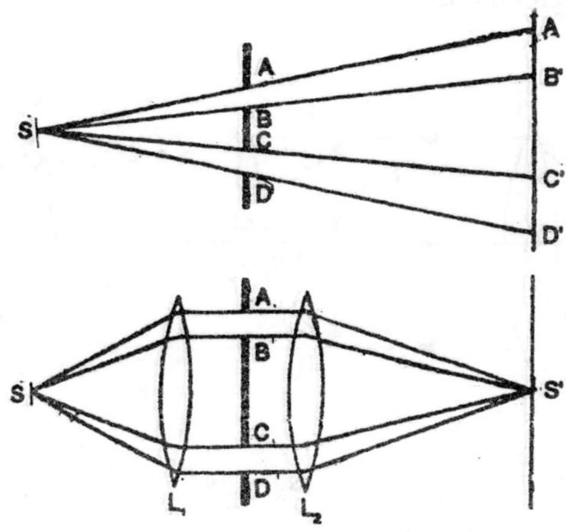

Fig. 3.27 : Fresnel and Fraunhofer diffractions.

In Fig. 3.27 (bottom), lenses L_1 and L_2 are added so that S' is an map of the source S. Now, the gross structure of S' has nothing to do with the structure of AB or CD, it is an image of S. The modifications in the detailed structure of S' due to apertures which limit the wavefront are called Fraunhofer diffraction, which we will now study.

It is of interest to note that if in Fig. 3.27 (lower) lens L_1 or L_2 or both are removed and correspondingly S or S' or both are shifted to infinite distance, the situation in effect does not change. For this reason,

Fraunhofer diffraction case is also defined as that in which the source as well as the plane of observation are at infinite distance from the diffracting element. As one result, the angular inclinations are important here, whereas lateral distances are important in the Fresnel class.

As we shall see, Fraunhofer class of diffraction can be treated in rigorous mathematical terms and has wide-ranging applications.

FRAUNHOFFER DIFFRACTION AT A SINGLE SLIT

In Fig. 3.28 S is a narrow slit perpendicular to the plane of the paper and illuminated by monochromatic light. L_1 is the collimating lens and AB is a slit of width a. XY is the incident spherical wave front. The light passing through the slit AB is incident on the lens L_2 and the final refracted beam is observed on the screen MN. The screen is perpendicular to the plane of the paper. The line SP is perpendicular to the screen. L_1 and L_2 are achromatic lenses.

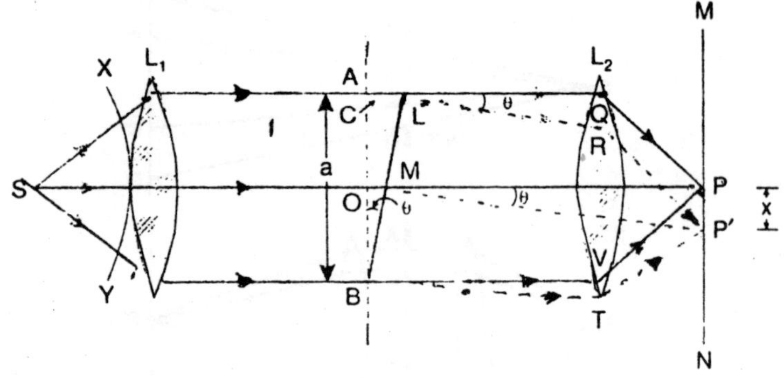

Fig. 3.28

A plane wave front is incident on the slit AB and each point on this wave front is a source of secondary disturbance. The secondary waves traveling in the direction parallel to OP viz. AQ and BV come to focus at P and a bright central image is observed. The secondary waves from points equidistant from O and situated in the upper and lower halves OA and OB of the wave front travel the same distance in reaching P and hence the path difference is zero. The secondary waves reinforce one another and P will be a point of maximum intensity.

Now, consider the secondary waves traveling in the direction AR, inclined at an angle θ to the direction OR All the secondary waves traveling in this direction reach the point P' on the screen. The point P' will be of maximum or minimum intensity depending on the path difference between the secondary waves originating from the corresponding points of the wave front. Let us draw OC and BL perpendicular to AR.

Then, in ΔABL

$$\sin\theta \frac{AL}{AB} = \frac{AL}{a}$$

or

$$AL = a\sin\theta \qquad ...(18)$$

where a is the width of the slit and AL is the path difference between the secondary waves originating from A and B. If this path difference is equal to λ the wavelength of the light used, then P' will be a point of minimum intensity. The whole wave front can be considered to be of two halves OA and OB and if the path difference between the secondary waves from A and B is λ then the path difference between the secondary waves from A and O will be $\lambda/2$. Similarly, for every point in the upper half OA, there is a corresponding point in the lower half OB, and the path difference between the secondary waves from these points is $\lambda/2$ .Thus, destructive interference takes place and the point P' will be of minimum intensity. If the direction of the secondary waves is such that AL = 2λ, then also the point where they meet the screen will be of minimum intensity. This is so because the secondary waves from the corresponding points of the lower half differ in path by $\lambda/2$. And this again gives the position of minimum intensity. In general,

$$a\sin\theta_n = n\lambda$$

$$\sin\theta_n = \frac{n\lambda}{a} \qquad ...(19)$$

where θ_n gives the direction of the nth minimum. Here n is an integer. If, however, the path difference is odd multiples of $\lambda/2$, the directions of the secondary maxima can be obtained. In this case,

$$a\sin\theta_n = (2n+1)\lambda/2$$

$$\sin\theta_n \frac{(2n+1)\lambda}{2a} \qquad ...(20)$$

where n = 1, 2, 3,...

Thus, the diffraction pattern due to a single slit consists of a central bright maximum at P followed by secondary maxima and minima on both the sides, as shown in Fig. 3.29. P corresponds to the position of the central bright maximum and the points on the screen for which the path difference between the points A and B is λ, 2λ etc correspond to the position of secondary minima. The secondary maxima are of much less intensity. The intensity falls off rapidly from the point P outwards.

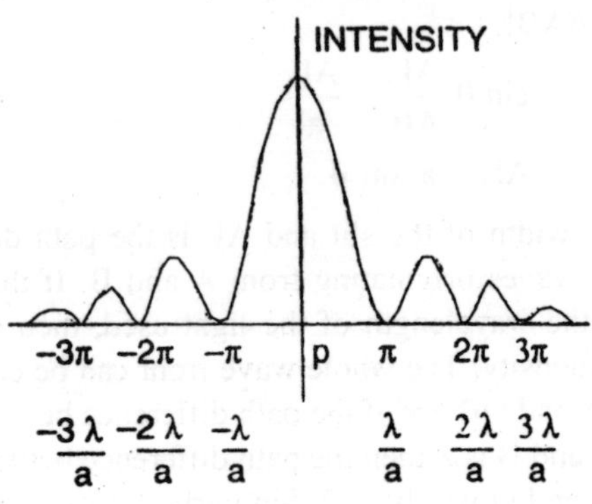

Fig. 3.29

If the lens L_2 is very near the slit or the screen is far away from the lens L–2 then,

$$\sin\theta_n \; \frac{x}{f} \qquad ...(21)$$

where f is the focal length of the lens L_2.

But $\sin\theta_n \; \frac{\lambda}{a}$ ...(22)

$$\therefore \qquad \frac{x}{f} = \frac{\lambda}{a} \quad \text{or} \quad x = \frac{f\lambda}{a}$$

where x is the distance of the secondary minimum from the point P.

Thus, the width of the central maximum W = 2x

or $$W = \frac{2f\lambda}{a} \qquad ...(23)$$

The width of the central maximum is proportional to the wavelength of the light. With red light (longer wavelength), the width of the central maximum is more than with violet light (shorter wavelength). With a narrow slit, the width of central maximum is more. The diffraction pattern consists of alternate bright and dark bands with monochromatic light. With white light, the central maximum is white and the rest of the diffraction bands are coloured. If the width a of the slit is large, $\sin\theta$ is small and hence θ is small. The maxima and minima are very close to the central maximum at P. But with a narrow slit, a is small and hence θ is large. This results in a distinct diffraction maxima and minima on both the sides of P.

Intensity Distribution in Diffraction Pattern Due to a Single Slit

The intensity variation in the diffraction pattern due to a single slit can be investigated as follows. The incident plane wave front on the slit AB (Fig. 3.30) can be imagined to be divided into a large number of infinitesimally small strips. The path difference between the secondary waves emanating from the extreme points A and B is a sin 9 where a is the width of the slit and ∠ABL = θ. For a parallel beam of incident light, the amplitude of vibration of the waves from each strip can be taken to be the same. As one considers the secondary waves in a direction inclined at an angle θ from the point B upwards, the path difference changes and hence the phase difference also increases. Let a be the phase difference between the secondary waves from the points B and A of the slit (Fig. 3.29). As the wave front is divided into a large number of strips, the resultant amplitude due to all the individual small strips can be obtained by the vector polygon method. Here, the amplitudes are small and the phase difference increases by infinitesimally small amounts from strip to strip.

Thus, vibration polygon coincides with the circular arc OM (Fig. 3.30). OP gives the direction of the initial vector and NM the direction of the final vector due to the secondary waves from A. K is the centre of the circular arc.

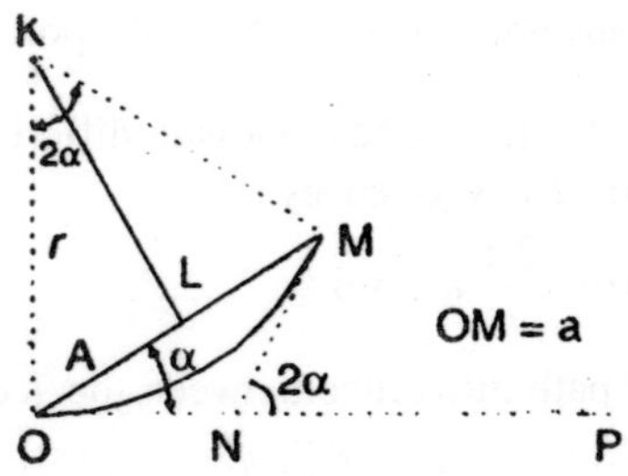

Fig. 3.30

$\angle MNP = 2\alpha \qquad \therefore\ ZOKM = 2\alpha$

In ΔOKL $\qquad \sin\alpha = \dfrac{OL}{r};$

$\therefore\ OL = r\sin\alpha$

where r is the radius of the circular arc.

$\therefore$ Chord OM = 2 OL = 2 r sin α ...(24)

The length of the arc OM is proportional to the width of the slit.

Length of the arc OM = Ka

where K is a constant and a is the width of the slit.

Also $$2\alpha = \frac{\text{Arc OM}}{\text{radius}} = \frac{Ka}{r}$$

or $$2r = \frac{Ka}{\alpha} \qquad ...(25)$$

Substituting the value of 2r in equation *24), we get

$$\text{Chord OM} = \frac{Ka}{\alpha}.\ \sin \alpha$$

But, OM = A where A is the amplitude of the resultant vibration.

$\therefore$ $$A = (Ka)\frac{\sin \alpha}{\alpha}$$

or $$A = A_0 \frac{\sin \alpha}{\alpha} \qquad ...(26)$$

Thus, the resultant amplitude of vibration at a point on the screen is given by An $A_0 \dfrac{\sin \alpha}{\alpha}$ and the intensity I at the point is given by

$$I = A^2 = A_0^2 \frac{\sin^2 \alpha}{\alpha^2} = I_0 \left(\frac{\sin \alpha}{\alpha}\right)^2 \qquad ...(27)$$

The intensity at any point on the screen is proportional to $\left(\dfrac{\sin \alpha}{\alpha}\right)^2$. A phase difference of 2n corresponds to a path difference of λ. Therefore, a phase difference of 2α is given by

$$2\alpha = \frac{2\pi}{\lambda} a \sin \theta \qquad ...(28)$$

where a sin θ is the path difference between the secondary waves from A and B (Fig. 7.2).

$$\alpha = \frac{\pi}{\lambda} a \sin \theta \qquad ...(29)$$

Thus, the value of α depends on the angle of the diffraction θ. The value of $\left(\dfrac{\sin^2 \alpha}{\alpha^2}\right)$ different values of θ gives the intensity at the point under consideration. Fig. 3.30 represents the intensity distribution. It is a graph of $\left(\frac{\sin^2 \alpha}{\alpha^2}\right)$ (along the Y-axis), as a function of α or $\sin \theta$ (along the X-axis).

DIFFRACTION PATTERN DUE TO A NARROW SLIT

S is a narrow slit illuminated by monochromatic light. The length of the slit is perpendicular to the plane of the paper. AB is a rectangular aperture parallel to the slit, MN is the screen and P is a point on the screen such that SOP is perpendicular to the plane of the paper; XY is the incident cylindrical wave front (Fig. 3.31). On the screen, EP is the illuminated portion and above E and below F is the region of the geometrical shadow.

If the slit AB is wide, then with reference to the point P, the cylindrical wave front can be divided into a large number of half period strips and the resultant amplitude at P will be $\frac{m_1}{2}$ where m_1 is the amplitude due to the first half period strip. Thus, the point P will be illuminated. Even points very near to P will be equally illuminated. If the wave front is divided with reference to points nearer P, the number of half period strips above and below the new pole in the exposed portion of the wavefront will be quite large and hence this results in uniform illumination.

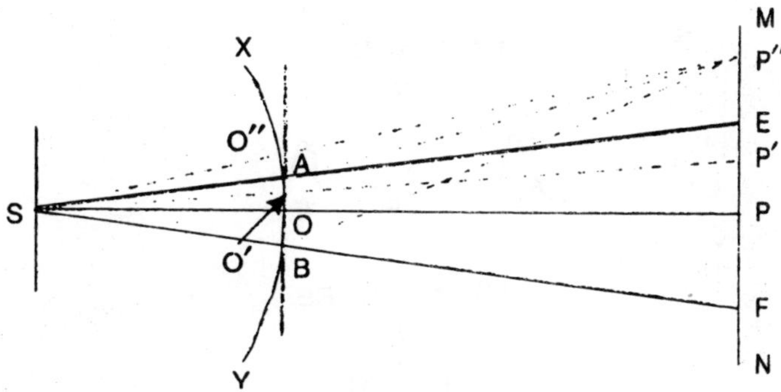

Fig. 3.31

Now, let us consider a point P' nearer to the edge of the geometrical shadow (Fig. 3.31). Let us join S to P'. Here O' is the pole of the wave front with reference to the point P'. If the wave front is divided into half period strips, the number of half period strips between O' and B will be quite large and the illumination at P' due to the lower portion of the wave front will be the same at all points near the edge of the geometrical shadow. But the intensity at P due to the exposed portion of the wave front between A and O' will depend on the number of half period strips present. If the number of half period elements is odd, the point P' will be of maximum intensity and if it is even the point will be of minimum intensity.

Let P" be a point in the region of the geometrical shadow. Let us join S to P". Here O" is the pole of the wave front with reference to the point P". If the wave front is divided into half period elements, then the upper half of the wave front between X and O" is cut off by the obstacle and only a portion between A and B is exposed to the point P". If the number of half period elements exposed by AB is odd, then P" will be of maximum intensity and if it is even, it will be of minimum intensity. But as the most effective central half period strips between O" and A are cut off, the intensity falls off rapidly in the region of the geometrical shadow and maxima and minima cannot be distinguished. The intensity distribution due to a wide aperture is shown in Fig. 3.32(b).

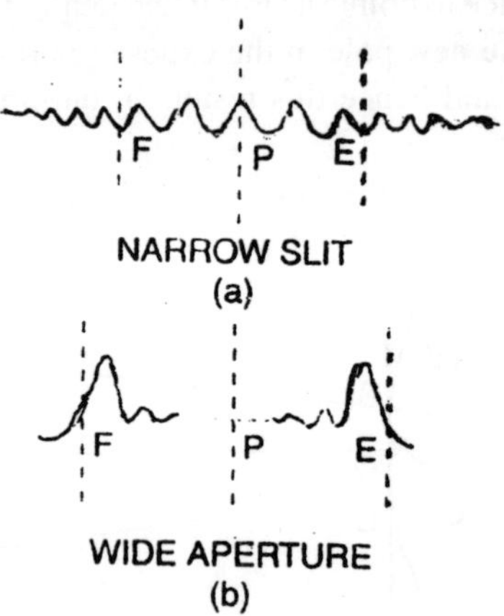

Fig. 3.32

On the other hand if the slit is narrow, the intensity at the point P will depend on the number of half period strips that can be constructed on the exposed wave front between A and B. If the number of half period strips is odd, the intensity at P will be maximum and if it is even, the intensity at P will be minimum (Fig. 3.32). Thus, the point P can be bright or dark. If we consider a point P' in the illuminated portion EF of the screen, the intensity at P will depend on number of half period strips that can be constructed between A and O' where O' is the pole of the wave front with reference to the point P'. If the number of half periods strips between A and O' is odd, P' will be a point of maximum intensity. Thus, between E and F alternate bright and dark bands will be observed and the point P may be bright or dark.

Now consider a point P" in the region of the geometrical shadow (Fig. 3.32). O" is the pole of the wavefront with reference to the point P" and the intensity at P" will depend on the number of half period strips exposed by the slit AB. The upper half of the wavefront above O" is obstructed by the obstacle and even the most effective central half period strips between O" and A are cut off by the obstacle. Thus, at P" which is far away from E, the maxima and minima become indistinguishable. There is no marked transition between the diffraction bands observed in the geometrical shadow and the illuminated portion. In the intensity distribution on the screen due to a narrow slit (say less than the wavelength of light), a broad central maximum will be observed in the illuminated portion and the intensity variation cannot be distinguished. The intensity gradually falls off in the region of geometrical shadow.

DIFFRACTION DUE TO A NARROW WIRE

In Fig. 3.33, S is a narrow slit illuminated by monochromatic light, AB is the diameter of the narrow wire and MN is the screen. The length of the wire is parallel to the illuminated slit and perpendicular to the plane of the paper.

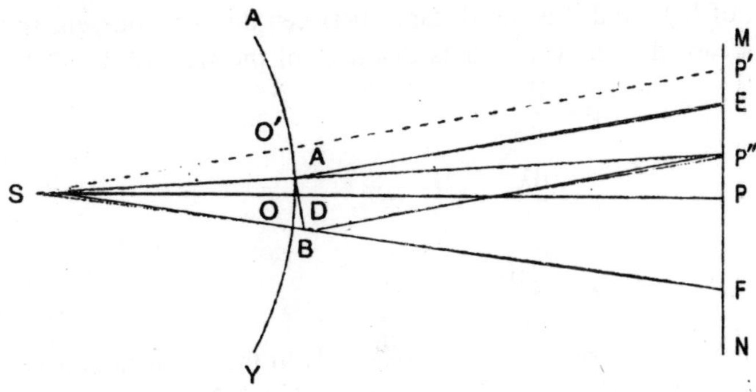

Fig. 3.33

The screen is also perpendicular to the plane of the paper. XY is the incident cylindrical wave front. P is a point on the screen such that SOP is perpendicular to the screen. EF is the region of the geometrical shadow and above E and below F, the screen is illuminated.

Now, let us consider a point P' on the screen in the illuminated portion. Let us join S to O', a point on the wave front. O' is the pole of the wave front with reference to P'. The intensity P' due to the wave front above O' is the same at all points and the effect due to the wave front BY is negligible. The intensity at P' will be a maximum or a minimum depending on whether the number of half period strips between O' and A is odd or even. Thus, in the illuminated portion of the screen/ diffraction bands of gradually diminishing intensity will be observed. The distinction between maxima and minima will become less if P' is far away from the edge E of the geometrical shadow. Maxima and minima cannot be distinguished if the wire is very narrow, because in that case the portion BY of the wavefront also produces illumination at P.

Next let us consider a point P" in the region of the geometrical shadow. Interference bands of equal width will be observed in this region due to the fact that the points A and B, of the incident wave front, are similar to two coherent-sources. The point P" will be of maximum or minimum intensity, depending on whether the path difference (BP" – AP") is equal to even or odd multiples of λ/2. The fringe width P is given by

$$\beta = \frac{d\lambda}{d}$$

where D is the distance between the wire and the screen, λ is the wave length of light and d is the distance between the two coherent sources. In this case d = 2r where 2r is diameter of the wire (AB = 2r).

$$\therefore \qquad \beta = \frac{d\lambda}{2r} \qquad (30)$$

$$\therefore \qquad r = \frac{d\lambda}{2\beta} \qquad ...(31)$$

$$\text{or} \qquad \lambda = \frac{2r\beta}{D} \qquad ...(32)$$

Here, β the fringe width corresponds to the distance between any two consecutive maxima. Thus, from equations (31) and (32), knowing the values of r or λ ; λ or r can be determined. In Fig. 3.34 the bands marked "a" represent the interference bands in the region of the geometrical shadow and the bands marked "b" and "c" represent the diffraction bands in the illuminated portion. The intensity distribution due to a narrow wire is shown in Fig. 3.33(a). The center of the geometrical shadow is bright.

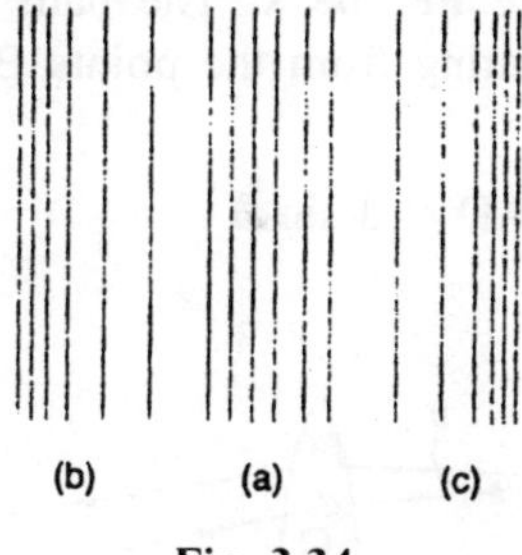

Fig. 3.34

On the other hand, if the wire is very thick, the interference bands cannot be noticed. From equation (19), $\beta = \frac{D\lambda}{2r}$; where β is the fringe width. As the diameter of the wire increases the fringe width decreases and if the wire is sufficiently thick, the width of the interference fringes decreases considerably and they cannot be distinguished. The intensity falls off rapidly in the geometrical shadow. The diffraction pattern in the illuminated portion will be similar to that of a thin wire Fig. 3.35. Coloured fringes will be observed with white light.

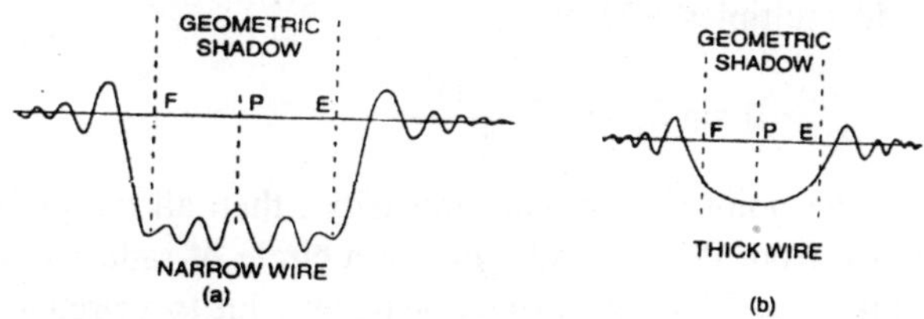

Fig. 3.35

FRAUNHOFFER DIFFRACTION AT A CIRCULAR APERTURE

In Fig. 3.36, AB is a circular aperture diameter d. C is the centre of the aperture and P is a point on the screen. CP is perpendicular to the screen. The screen is perpendicular to the plane of the paper. A plane wave front is incident on the circular aperture. The secondary wave traveling in the direction CO comes to the focus at P. Therefore, P corresponds to the position of the central maximum. Here, all the secondary waves emanating from points equidistant from O travel the same distance before reaching P and hence they all reinforce one another. Now, let us consider the secondary waves traveling in a direction inclined at an angle 6 with the direction CP. All these secondary waves meet at P_1 on the screen. Let the distance PP_1 be x. The path difference between the secondary waves emanating from the points B and A (extremities of diameter) is AD.

From the ΔABD, AD = d sin θ

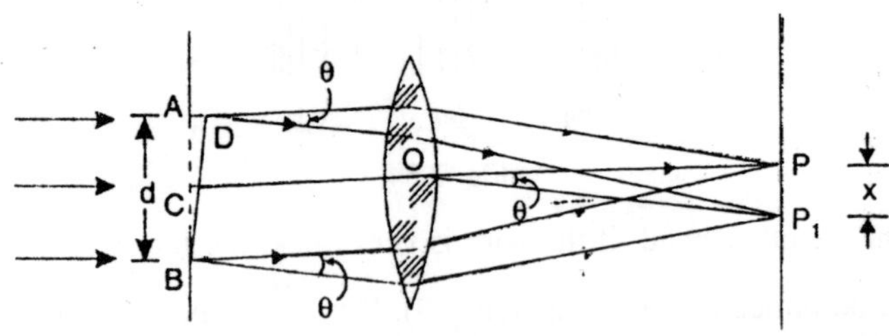

Fig. 3.36

The point P_1 will be of minimum intensity if this path difference is equal to integral multiples of λ, *i.e.*,

$$d \sin \theta_n = n\lambda, \qquad ...(33)$$

The point P_1 will be of maximum intensity if the path difference is equal to odd multiples of λ/2 *i.e.*,

$$d \sin \theta_n = \frac{(2n+1)\lambda}{2} \qquad ...(34)$$

If P_1 is the point of minimum intensity , then all the points at the same distance from P as P_1 and lying on a circle of radius x will be of minimum intensity. Thus, the diffraction pattern due to a circular aperture consists of a central disc called the Airy's disc, surrounded by alternate dark and bright concentric rings called the Airy's rings. The intensity of the dark rings is zero and that of the bright rings decreases gradually outwards from P. Further, if the collecting lens is very near the slit or when the screen is at a large distance from the lens,

$$\sin \theta = \theta = \frac{x}{f} \qquad ...(35)$$

Also, for the first secondary minimum, d sin θ = λ

$$\sin \theta = \theta = \frac{\lambda}{d} \qquad ...(36)$$

From equation (35) and (36)

$$\text{or} \qquad \frac{x}{f} = \frac{\lambda}{d} \qquad ...(37)$$

where x is the radius of the Airy's disc. But actually, the radius of the first dark ring is slightly more than that given by equation (17). According to Airy, it is given by

$$x = 1.22\frac{f\lambda}{d} \qquad ...(38)$$

The discussion on the intensity distribution of the bright and dark rings is similar to the one given for a rectangular slit. With increase in diameter of the aperture, the radius of the central bright ring decreases.

PLANE DIFFRACTION GRATING

A diffraction grating is an extremely useful device and in one of its forms it consists of a very large number of narrow slits side by side. The slits are separated by opaque spaces. When a wave front is incident on a grating surface, light is transmitted through the slits and obstructed by the opaque portions. Such a grating is called a *transmission grating.* The secondary waves from the positions of the slits interfere with one another, similar to the interference of waves in Young's experiment. Joseph Fraunhoffer used the first grating which consisted of a large number of parallel fine wires stretched on a frame. Now, gratings are prepared by ruling equidistant parallel lines on a glass surface. The lines are drawn with a fine diamond point. The space in between any two lines is transparent to light and the lined portion is opaque to light. Such surfaces act as transmission gratings. If, on the other hand, the lines are drawn on a silvered surface (plane or concave) then light is reflected from the positions of the mirror in between any two lines and such surfaces act as *reflection gratings*.

If the spacing between the lines is of the order of the wavelength of light, then an appreciable deviation of the light is produced. Gratings used for the study of the visible region of the spectrum contain 10,000 lines per cm. Gratings, with originally ruled surfaces are only few. For practical purposes, replicas of the original grating are prepared. On the original grating surface a thin layer of collodion solution is poured and the solution is allowed to harden. Then, the film of collodion is removed from the grating surface and then fixed between two glass plates. This serves as a plane transmission grating. A large number of replicas are prepared in this way from a single original ruled surface.

Theory of Plane Transmission Grating

In Fig. 3.37, XY is the grating surface and MN is the screen, both perpendicular to the plane of the paper. The slits are all parallel to one another and perpendicular to the plane of the paper. Here

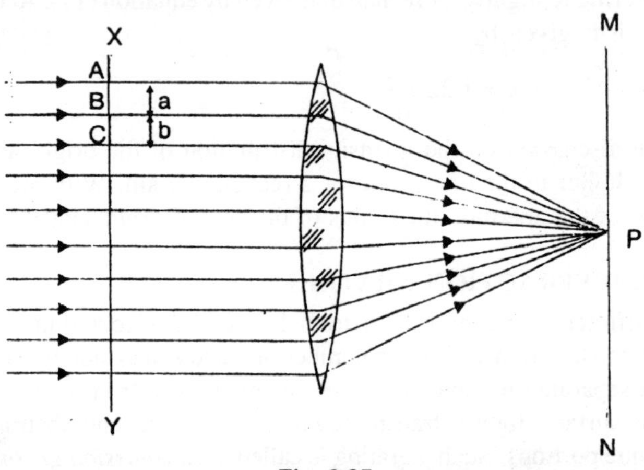

Fig. 3.37

AB is a slit and BC is an opaque portion. The width of each slit is a and the opaque spacing between any two consecutive slits is b. Let a plane wave front be incident on the grating surface. Then all the secondary waves traveling in the same direction as that of the incident light will come to focus at the point P on the screen. The screen is placed at the focal plane of the collecting lens. The point P where all the secondary waves reinforce one another corresponds to the position of the central bright maximum.

Now, consider the secondary waves traveling in a direction inclined at an angle θ with the direction of the incident light (Fig. 3.38). The collecting lens also is suitably rotated such that the axis of the lens is parallel to the direction of the secondary waves. These secondary waves come to focus at point P_1 on the screen.

The intensity at P_1 will depend on the path-difference between the secondary waves originating from the corresponding points A and C of two neighbouring slits. In Fig. 3.38, AB = a and BC = b. The path difference between the secondary waves starting from A and C is equal to AC sin θ.

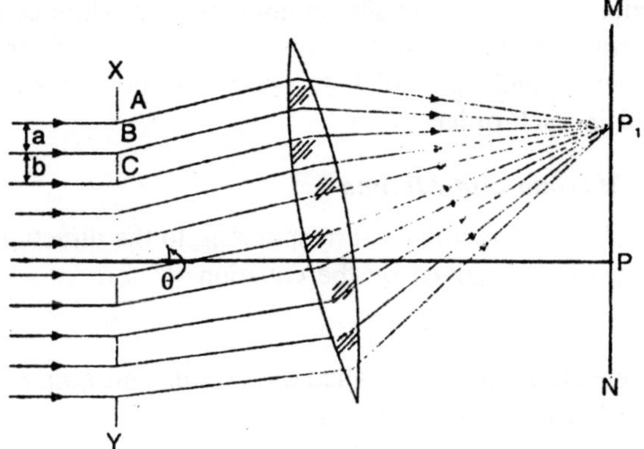

Fig. 3.38

But $$AC = AB + BC = a + b$$

Path difference = AC sin θ = (a + b) sin θ

The point P_1 will be of maximum intensity if this path difference is equal to integral multiples of λ where λ is the wavelength of light. In this case, all the secondary waves originating from the corresponding points of the neighbouring slits reinforce one another and the angle θ gives the direction of maximum intensity. In general

$$(a + b) \sin \theta = n\lambda \quad ...(39)$$

where θ_n is the direction of the nth principal maximum. Putting n = 1, 2, 3, etc, the angles θ_1, θ_2, θ_3, etc corresponding to the directions of the principal maxima can be obtained.

If the incident light consists of more than one wavelength, the beam gets dispersed and the angles of diffraction for different wavelengths will be different. Let λ and λ + dλ be two nearby wavelengths present in the incident light and θ and (θ + dθ) be the angles of diffraction corresponding to these two wavelengths. Then, for the first order principal maxima

$$(a + b) \sin \theta = \lambda$$

and $(a + b) \sin (\theta + d\theta) = \lambda + d\lambda$

Thus, in any order, the number of principal maxima corresponds to the number of wavelengths present. A number of parallel slit images corresponding to the different wavelengths will be observed on the screen, n = 1 gives the direction of the first order image, n= 1 gives the direction of the second order image and so on. When white light is used, the diffraction pattern on the screen consists of a white central bright maximum and on both sides of this maximum a spectrum corresponding to the different wavelengths of light present in the incident beam will be observed in each order.

Secondary Maxima and Minima

The angle of diffraction θ corresponding to the direction of the nth principal maximum is given by the equation

$$(a + b) \sin \theta_n = n\lambda$$

In this equation, (a + b) is called the *grating constant.* For a grating with 15,000 lines/inch, the value of

$$(a+b) = \frac{2.54}{15000} \text{ cm}$$

Now, let the angle of diffraction be increased by a small amount dθ such that the path difference between the secondary waves from the points A and C increases by λ/N (Fig. 3.38). Here N is the total number of lines on the grating surface. Then, the path difference between the secondary waves from the extreme points of the grating surface will be (λ/N)N = λ. Assuming the whole wave front to be divided into two halves, the path difference between the corresponding points of the two halves will be θ/2 and all the secondary waves cancel one another's effect. Thus, (θ_n+ dθ) will give the direction of the first secondary minimum after the nth primary maximum. Similarly, if the path difference between the secondary waves from the points A and C is 2λ/ N, 3λ/N etc, for gradually increasing values of dθ, these angles correspond to the directions of 2nd, 3rd etc secondary minima after the nth primary maximum.

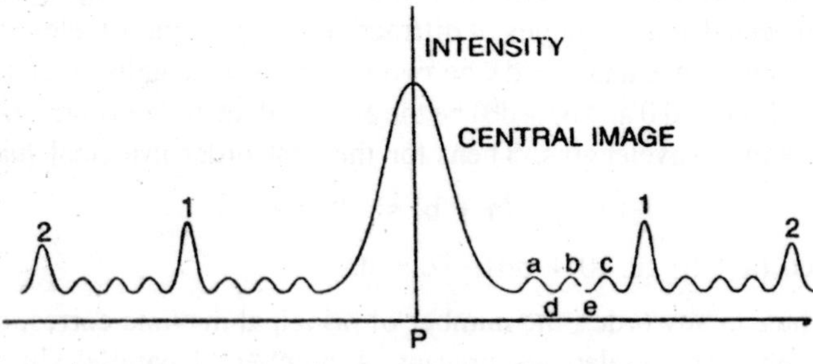

Fig. 3.39

If the value is 2λ/N, then the path difference between the secondary waves from the extreme points of the grating surface is

$(2\lambda/N)N = 2\lambda$ and considering the wave front to be divided into 4 portions, the concept of the 2nd secondary minimum can be understood. The number of secondary minima in between any two primary maxima is (N–l) and the number of secondary maxima is (N – 2).

The intensity distribution on the screen is shown in Fig. 3.40. P corresponds to the position of the central maxima and 1, 2, etc, on the two sides of P represent the 1st, 2nd, etc principal maxima. a, b, etc are secondary maxima and d, e etc are the secondary minima. The intensity as well as angular spacing of the secondary maxima and minima is so small in comparison to the principal maxima that they cannot be observed. It results in uniform darkness between any two principal maxima.

CORNU'S SPIRAL

To find the effect at a point due to an incident wave front Fresnel's method consists in dividing the wavefront into half period strips or half period zones. The path difference between secondary waves from two corresponding points of neighbouring zones is equal to $\lambda/2$.

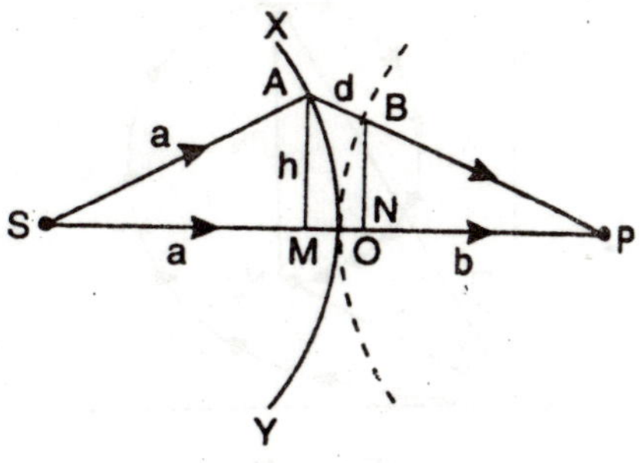

Fig. 3.40

In Fig. 3.40, S is a point source of light and XV is the incident spherical wave front. With reference to the point P, 0 is the pole of the wave front. Let a and b be the distances of the points S and P from the pole of the wave front.

With P as centre and radius b, let us draw a sphere touching the incident wavefront at O. The path difference between the waves travelling in the directions SAP and SOP is given by

$$d = SA + AP - SOP$$

$$= SA + AP - (SO + OP)$$

$$= a + AB + b - (a + b) = AB.$$

For large distances of a and b, AM and BN can be taken to be approximately equal and the path difference d can be written as

$$d = AB = MO + ON$$

But, from the property of a circle,

$$MO = \frac{AM^2}{2SO} = \frac{h^2}{2a} \text{ and } ON = \frac{BN^2}{2OP} = \frac{h^2}{2b} \text{ (approximately)}$$

$$\therefore \quad d = \frac{h^2}{2a} + \frac{h^2}{2b} = \frac{h^2}{2ab}(a+b) \qquad \text{...(40)}$$

If AM happens to be the radius of the nth half period zone, then this path difference is equal to $\frac{n\lambda}{2}$ according to the Fresnel's method of constructing the half period zones.

$$\therefore \qquad \frac{h^2}{2ab}(a+b) = \frac{n\lambda}{2} \qquad \text{...(41)}$$

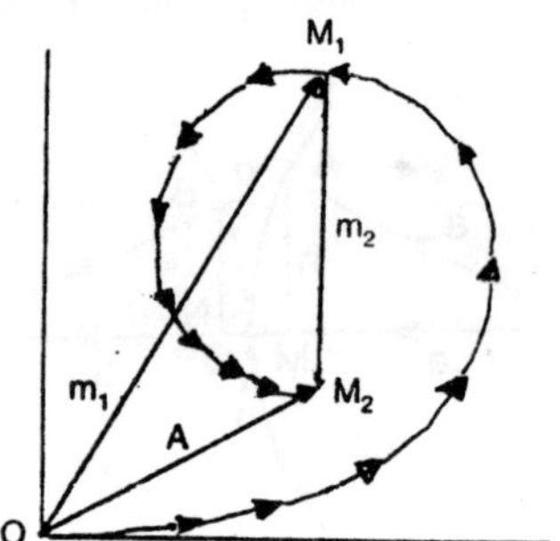

Fig. 3.41

The resultant amplitude at an external point due to the wave front can be obtained by the following method. Let the first half period strip of the Fresnel's zones be divided into eight sub- strips and these vectors are represented from O to M_1 (Fig. 3.41). The continuous phase change is due to the continuous increase in the obliquity factor from O to M_1. The resultant amplitude at the external point due to the first half period strip is given by OM_1 (= m_1).

Similarly, if the process is continued, we obtain the vibration curve

$$\frac{h^2}{2ab}(a+b) = \frac{n\lambda}{2}$$

The resultant amplitude at the point due to the first two half period strips is given by OM_2 (= A). If instead of eight sub-strips, each period zone is divided into sub-strips of infinitesimal width, a smooth curve will be obtained. The complete vibration curve for whole wave front will be a spiral as shown in Fig. 3.42. X and Y correspond to the two extremities of the wave front and M_1 and M_2 etc. refer to the edge of the first, second, etc. half period strips. Similarly M_1' and M_2' etc. refer to the edge of the first, second, etc. half period strips of the lower portion of the wave front. This is called *Cornu's spiral*. The characteristic of this curve is that for any point P on the curve, the phase lag δ is directly proportional to the square of the distance υ. The distance is measured along the length of the curve from the point O. For a path difference of λ, the phase difference Iv.. Hence, for a path difference of d, the phase difference δ is given by,

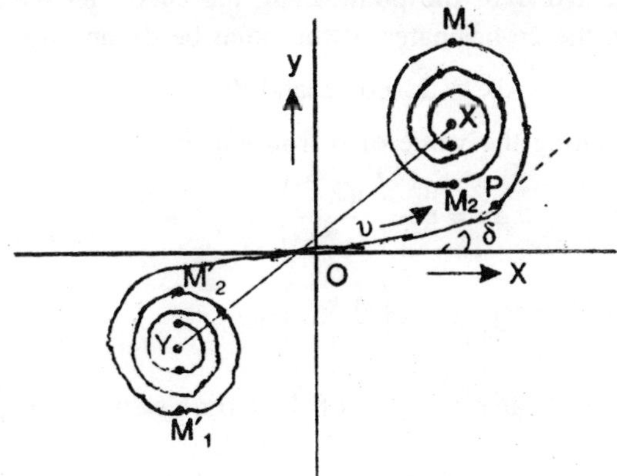

Fig. 3.42

$$\delta = \frac{2\pi}{\lambda} d.$$

Substituting the value of d from equation (40) we get,

$$\delta = \frac{\pi}{2}\left[\frac{2h^2\ (a+b)}{ab\lambda}\right] \quad ...(42)$$

$$\delta = \frac{\pi}{2}\upsilon^2 \quad ...(43)$$

The value of (i) is given by

$$\upsilon^2 = \frac{2h^2\ (a+b)}{ab\lambda}$$

or $$\upsilon = h\sqrt{\frac{2(a+b)}{ab\lambda}} \quad ...(44)$$

Cornu's spiral can be used for any diffraction problem irrespective of the values of a, b and λ.

Fresnel's Integrals

For any point on the Cornu's spiral, the x and y co-ordinates are given by two integrals known as *Fresnel's integrals*. Let us consider the point P on the spiral. The distance of the point P along the curve from the origin is υ. The tangent to the curve at P makes an angle δ with the x-axis, δ corresponds to the phase change from O to P. For a small displacement d υ of the point along the curve, let the corresponding changes in the co-ordinates of the point be dx and dy.

Then, $$dx = d\upsilon \cos \delta \text{ and } dy = d\upsilon \sin \delta$$

Substituting the value of δ from equation (43), we get

$$dx = \cos\left(\frac{\pi \upsilon^2}{2}\right) d\upsilon \quad ...(45)$$

and $$dy = \sin\left(\frac{\pi \upsilon^2}{2}\right) d\upsilon \quad ...(46)$$

The coordinates x and y of the Cornu's spiral are given by,

$$x = \int dx = \int_0^{\upsilon} \cos\left(\frac{\pi \upsilon^2}{2}\right) d\upsilon \quad ...(47)$$

and $$y = \int dx = \int_0^{\upsilon} \sin\left(\frac{\pi \upsilon^2}{2}\right) d\upsilon \quad ...(48)$$

These are called *Fresnel's Integrals*

Maxima and Minima in Diffraction Patterns (Cornu's Spiral)

The various diffraction patterns as discussed in the earlier articles and the positions of maxima and minima can be easily explained with the help of Cornu's spiral.

In Fig. 3.43, O is the origin of coordinates, OX is the vibration curve for the upper half of the wave front and OY refers to the vibration curve for the lower half of the wave front. If the whole wave front is unobstructed, the resultant amplitude at a point is given by XY.

If a cylindrical wave front is incident on a straight edge, the amplitude at a point P on the edge of the geometrical shadow (refer to the discussion on diffraction at a straight edge) is given by XY. If points above P in the illuminated portion are considered, gradually more of the lower half of the wave front is also exposed to the screen and amplitude vector passes through maxima and minima. Xb', Xd' etc. refer to maximum amplitudes and Xc' refers to the minimum amplitude. Thus, in the illuminated portion alternate bright and dark bands parallel to the length of the slit are observed on the screen. If points below P and in the region of (the geometrical shadow are considered) the lower half of the wave front and a portion of the upper half of the wave front are cut off and the tail of the amplitude vector moves to the right of O. The amplitude gradually decreases and becomes zero when the tail approaches X. Thus, in the region of the geometrical shadow the intensity falls off gradually. Quantitative values of intensity for different points on the screen can be obtained by finding the amplitude A for different values of υ.

The square of the amplitude measures the intensity at that point. The point M_1, M_2 etc correspond to the edges of the first, second etc half period strips in the upper half of the wave front and the points M_1', M_2' etc refer to the lower half of the wave front. The points b', d' etc on the spiral, corresponding to maximum intensity, occur a little before the points M_1', M_3' etc are reached.

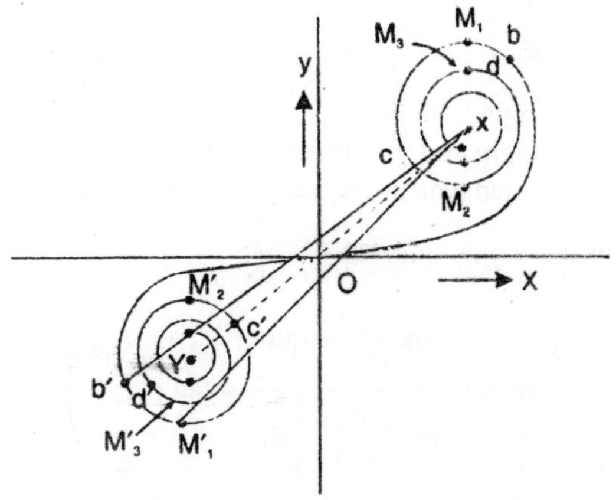

Fig. 3.43

The coordinates x and y of the Cornu's spiral are given by

$$x = \int dx = \int_0^{\upsilon} \cos\left(\frac{\pi\upsilon^2}{2}\right) d\upsilon \qquad ...(49)$$

$$y = \int dy = \int_0^{\upsilon} \cos\left(\frac{\pi\upsilon^2}{2}\right) d\upsilon \qquad ...(50)$$

The values of these integrals can be calculated for different values ef υ. The graph is as shown in Fig. 3.43. The two integrals represent the horizontal and vertical components of the resultant amplitude. The intensity is proportional to the square of the resultant amplitude.

$$I_P = k\left[x^2 + y^2\right] \qquad ...(51)$$

When the whole of the wave front is exposed to the point, $\upsilon \to \infty$ the values of the integrals will be,

$$x = \int_0^{\infty} \cos\left(\frac{\pi\upsilon^2}{2}\right) d\upsilon = \frac{1}{2}$$

$$y = \int_0^{\infty} \sin\left(\frac{\pi\upsilon^2}{2}\right) d\upsilon = \frac{1}{2}$$

Thus, for the point X in Fig 3.43, the x and y co-ordinates are $\left[\frac{1}{2}, \frac{1}{2}\right]$. Similarly, for the point Y on the lower half of the spiral, the coordinates are $\left[-\frac{1}{2}, -\frac{1}{2}\right]$.

At the origin, *i.e.* when $\upsilon = 0$, $x = 0$ and $y = 0$, the spiral passes through the origin and it is symmetrical with the origin. At any point on the spiral, the tangent to the curve makes an angle ϕ

with the x-axis and $\tan\phi = \frac{dy}{dx}$

$$\tan\phi = \frac{\sin\left(\frac{\pi\upsilon^2}{2}\right) d\upsilon}{\cos\left(\frac{\pi\upsilon^2}{2}\right) d\upsilon} = \tan\left(\frac{\pi\upsilon^2}{2}\right)$$

or $$\phi = \left(\frac{\pi\upsilon^2}{2}\right) \qquad ...(52)$$

When $\upsilon = 0, \quad \phi = 0.$

It means the curve is parallel to the x-axis at the origin.

The element of length $d\upsilon$ along the spiral is given by,

$$d\upsilon = \sqrt{(dx)^2 + (dy)^2} \quad ...(53)$$

Differentiating equation (52), we get

$$d\phi = \frac{2\pi\upsilon}{2} d\upsilon$$

or

$$\frac{d\upsilon}{d\phi} = \frac{1}{\pi\upsilon} \quad ...(54)$$

Hence $\frac{d\upsilon}{d\phi}$ measures the radius of curvature of the spiral at the point under consideration.

From equation (53), $\frac{d\upsilon}{d\phi} \propto \frac{1}{\upsilon}$

It shows that with the increase in the value of υ, the radius of curvature of the curve gradually decreases and takes the shape of a spiral. Finally for $\upsilon \to \infty$, the curve ends in a point (X or Y).

WIDTH OF PRINCIPAL MAXIMA

The direction of the nth principal maximum is given by

$$(a + b) \sin \theta_n = n\lambda$$

Let $\theta_n + d\theta$ and $\theta - d\theta$ give the directions of the first secondary minima on the two sides of the nth primary maxima (Fig. 3.44). Then

$$(a + b) \sin [\theta_n - d\theta] = n\lambda \pm \frac{\lambda}{N} \quad ...(55)$$

where N is the total number of the lines on the grating surface.

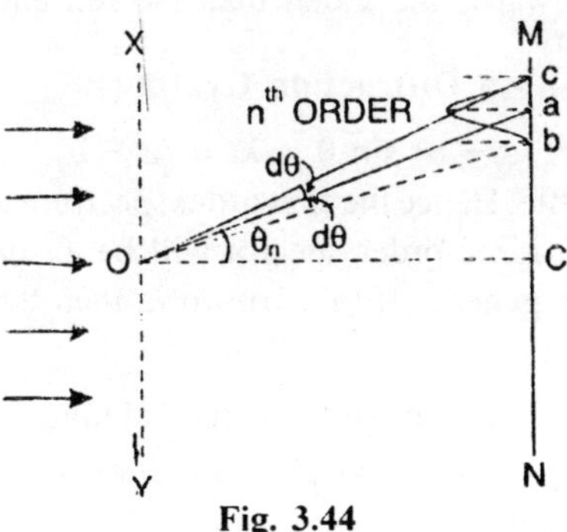

Fig. 3.44

Dividing (43) by (42), we get

$$\frac{(a+b)\sin[\theta_n \pm d\theta]}{(a+b)\sin\theta_n} = \frac{n\lambda \pm \frac{\lambda}{N}}{n\lambda}$$

$$\frac{\sin[\theta_n \pm d\theta]}{\sin\theta_n} = 1 \pm \frac{1}{nN}$$

Expanding this equation, we get

$$\frac{\sin\theta_n.\cos d\theta \pm \cos\theta_n \sin d\theta]}{\sin\theta_n} = 1 \pm \frac{1}{nN} \qquad \text{...(56)}$$

For small values of $d\theta$; $\cos d\theta = 1$ and $\sin d\theta = d\theta$.

$$\therefore \qquad 1 \pm \cot\theta_n\, d\theta = 1\ \frac{1}{nN}$$

or $$\cot\theta_n\, d\theta = \frac{1}{nN}$$

$$\therefore \qquad d\theta = \frac{1}{nN\cos\theta_n} \qquad \text{...(57)}$$

In equation (54), $d\theta$ refers to half the angular width of the principal maximum. The half width $d\theta$ is (i) inversely proportional to N, the total number of lines and (ii) inversely proportional to n cot θ_n. The value of n cot θ_n is more for higher orders because the increase in the value of cot θ_n is less than the increase in the order. Thus, the half width of the principal maximum is less for higher orders. Also, the larger the number of lines on the grating surface, the smaller is the value of $d\theta$. Further, the value of θ_n is higher for longer wavelengths and hence the spectral lines are sharper towards the violet than the red end of the spectrum.

Absent Spectra with a Diffraction Grating

In the equation $(a + b)\sin\theta = \lambda$, if $(a + b) < \lambda$, then $\sin\theta > 1$. But this is not possible. Hence the first order spectrum is absent. Similarly, the second, the third, etc order spectra will be absent if $(a + b) < 2\lambda$, $(a + b) < 3\lambda$ etc. In general, if $(a + b) < n\lambda$, then the n order spectrum will be absent.

The condition for absent spectra can be obtained from the following considerations. For the nth order principal maximum

$$(a + b)\sin\theta_n = n\lambda \qquad \text{...(58)}$$

Further, if the value of a and θ_n are such that

$$a \text{ sm } \theta_n = \lambda \quad ...(59)$$

then, the effect of the wave front from any particular slit will be zero. Considering each slit to be made up of two halves, the path difference between the secondary waves from the corresponding points will be λ/2 and they cancel one another's effect. If the two conditions given by equations (25) and (26) are simultaneously satisfied, then dividing (18.51) by (26), we get

$$\frac{(a+b)\sin\theta_n}{a \sin\theta_n} = \frac{n\lambda}{\lambda}$$

or

$$\frac{a+b}{a} = n \quad ...(60)$$

same line. The position of the telescope is noted on the circular scale and 90° is added to this reading. The telescope is turned to this position. In this position the axis of the telescope is perpendicular to the axis of collimator. The position of the telescope is fixed. The given transmission grating is mounted at the centre of the prism table such that the grating surface is perpendicular to the prism table. The prism table is suitably rotated such that the image of the slit reflected from the grating surface is obtained in the centre of field of view of the telescope. This means that the parallel rays of light from the collimator are incident at an angle 45° on the grating surface because the axis of the collimator and the telescope are perpendicular to each other. The reading of the prism table is noted and adding 45° to this reading, the prism table is suitably rotated to the new position so that the grating surface is normal to the incident light.

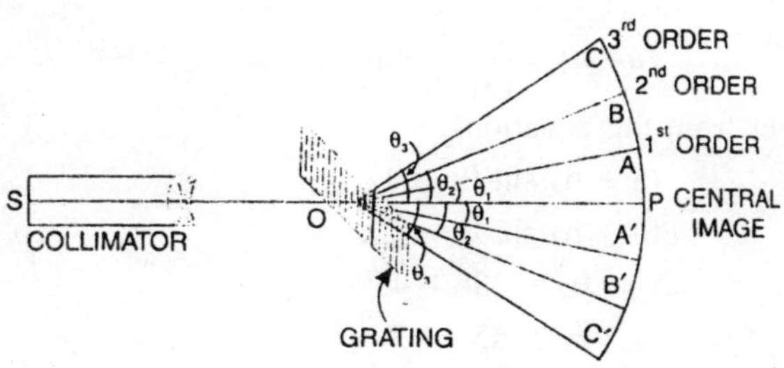

Fig. 3.45

If the wavelength of the sodium light is to be determined, then the angles of diffraction θ_1 and θ_2 corresponding to the first and second order principal maxima are determined (Fig. 3.45). OA, OB etc., give the directions of the telescope corresponding to the first and second order images. A', B' etc refer to the positions of these images towards the left of the central maximum. The angles AOA' and BOB' are measured and half of these angles measure θ_1 and θ_2. Then

$$(a + b) \sin \theta_1 = 1\lambda \quad ...(61)$$

$$\text{and } (a + b) \sin \theta_2 = 2\lambda \quad ...(62)$$

Then the value of λ is calculated from equations (61) and (62) and the mean value is taken, (a + b) is the grating element and it is equal to the reciprocal of the number of lines per cm. If the number of lines on the grating surface is 15,000 per inch then

$$(a+b) = \frac{2.54}{15000} \text{ cm}$$

If the source of light emits radiations of different wavelengths, then the beam gets dispersed by the grating and in each order a spectrum of the constituent wavelengths is observed. To find the wavelength of any spectral line, the diffracting angles are noted in the first and second orders and using the equations given above, the wavelength of the spectral line can be calculated. Overlapping spectral orders can be avoided by using suitable colour filters so that the wavelengths beyond the range of study are eliminated.

With a diffraction grating, the wavelength of the spectral line can be determined very accurately. The method involves only the accurate measurement of the angles of diffraction.

Taking $\lambda = 6000 \text{ Å} = 6000 \times 10^{-8} \text{ cm}$

and $(a+b) = \dfrac{2.54}{15000} \text{ cm}$

We get from the equations,

$$(a + b) \sin \theta_1 = 1\lambda$$

$$(a + b) \sin \theta_1 = 2\lambda$$

and $\theta_1 = 20° - 45'$

and $\theta_2 = 45° - 7'$

As the angles are large they can be measured accurately with a properly calibrated spectrometer. The number of lines per inch (or cm), is given on the grating by the manufacturing company and hence (a + b) can be calculated. As the method does not involve measurements of very small distances (as in the case of interference experiments) an accurate value of λ can be obtained.

Dispersive Power of Grating

Dispersive power of a grating is defined as the ratio of the difference in the angle of diffraction of any two neighbouring spectral lines to the difference in wavelength between the two spectral lines. It can also be defined as the difference in the angle of diffraction per unit change in wavelength. The diffraction of the nth order principal maximum for a wavelength λ is given by the equation

$$(a + b) \sin \theta = n\lambda$$

Differentiating this equation with respect to θ and λ, we get

$$(a + b) \cos \theta \, d\theta = n d\lambda$$

or
$$\frac{d\theta}{d\lambda} = \frac{n}{(a+b)\cos\theta} = \frac{nN'}{\cos\theta} \quad ...(63)$$

From equ. (63) it is clear that the dispersive power of the grating is (i) directly proportional to the order of the spectrum, n (ii) directly proportional to the number of lines per cm, N' and (in) inversely proportional to cos θ. Thus, the angular spacing of any two spectral lines is double in the second order spectrum than that in the first order. Secondly, the angular dispersion of the lines is more with a grating having a larger number of lines per cm. Thirdly, the angular dispersion is a minimum when θ = 0. If the value of θ is not large, the value of cos θ can be taken as unity and the influence of this factor can be neglected. Then it is clear that the angular dispersion of any two spectral lines is directly proportional to the difference in wavelength of the spectral lines. A spectrum of this type is called a *normal spectrum.*

If the linear spacing of two spectral lines of wavelengths λ and λ + dλ is dx in the focal plane of the telescope objective or photographic plate, then

$$dx = f \, d\theta$$

where f is the focal length of the objective. The linear dispersion is

$$\frac{dx}{d\lambda} = f\frac{d\theta}{d\lambda} = \frac{fnN'}{\cos\theta} \quad ...(64)$$

or
$$dx = \frac{fnN'}{\cos\theta}.d\lambda$$

The linear dispersion is useful in studying the photographs of a spectrum.

Prism and Grating Spectra

For dispersing a given beam of light and for studying the resultant spectrum, a diffraction grating is mostly used instead of a prism. The grating and prism spectra differ in the following points.

(i) With a grating, a number of spectra of different orders can be obtained on the two sides of the central maximum whereas with a prism only one spectrum is obtained.

(ii) The spectra obtained with a grating are comparatively purer than those with a prism.

(iii) Knowing the grating element (a + b) and measuring the diffraction angle, the wavelength of any spectral line can be measured accurately. But in case of a prism, the angles of deviation are not directly related to the wavelength of the special line. The angles of deviation are dependent on the refractive index of the material of the prism, which depends on the wavelength of light.

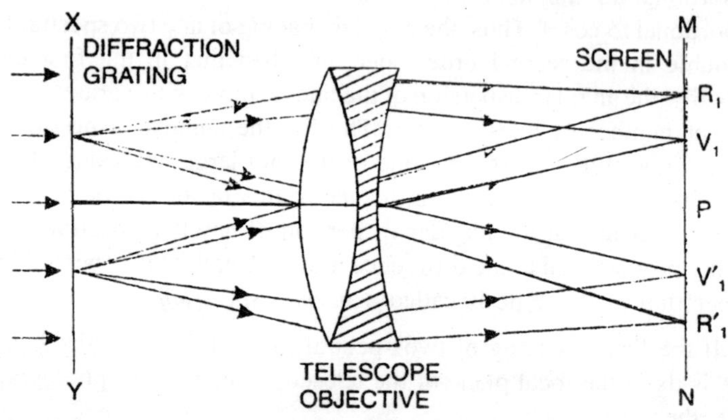

Fig. 3.46

(iv) The dispersive power of a grating is given by

$$\frac{d\theta}{d\lambda} = \frac{nN'}{\cos\theta}$$

and this is constant for a particular order. Thus, the spectral lines are evenly distribute. Hence, the spectrum obtained with a grating is said to be *rational* (Fig. 3.47). The refractive index of the material of a prism changes more rapidly at the violet end than at the red end of the Spectrum. The dispersive power of a prism is given by $\frac{d\mu}{\mu - 1}$, and this has higher value in the violet region than in the red region. Hence, there will be more spreading of the spectral lines towards the violet and the spectrum obtained with a prism is said to be *irrational* (Fig 3.47).

(v) The resolving power of a grating is given by nN whereas the resolving power of a prism given by $t \frac{d\mu}{d\lambda}$ where t is the base of the prism. The resolving power of a grating is much higher than that of a prism. Hence the same two nearby spectral lines appear better resolved with a grating than with a prism.

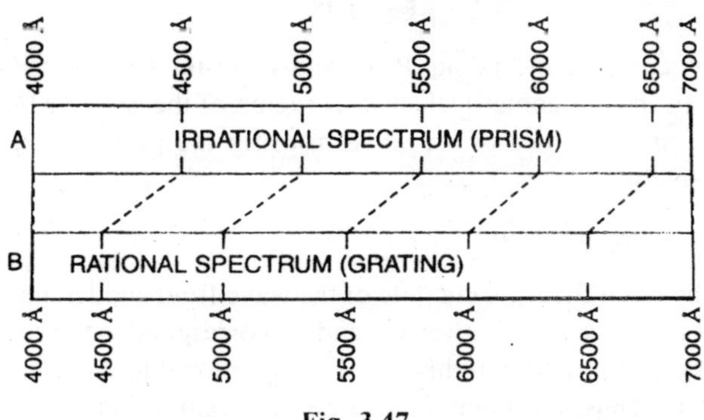

Fig. 3.47

(vi) Lastly, the spectra obtained with different gratings are identical because the dispersive power and the resolving power of a grating do not depend on the nature of the material of the grating. But the spectra obtained with prisms made of different materials are never identical because both dispersive and resolving powers depend on the nature of the material of the prism.

DIFFRACTION AT A STRAIGHT EDGE

S is a narrow slit illuminated by monochromatic light of wavelength λ. The length of the slit is perpendicular to the plane of the paper. AD is a straight edge and the length of the edge is parallel length of the slit (Fig. 3.48). AT is the incident cylindrical wave front. Cornu's spiral helps to obtain qualitatively the intensity distribution at a point on the screen.

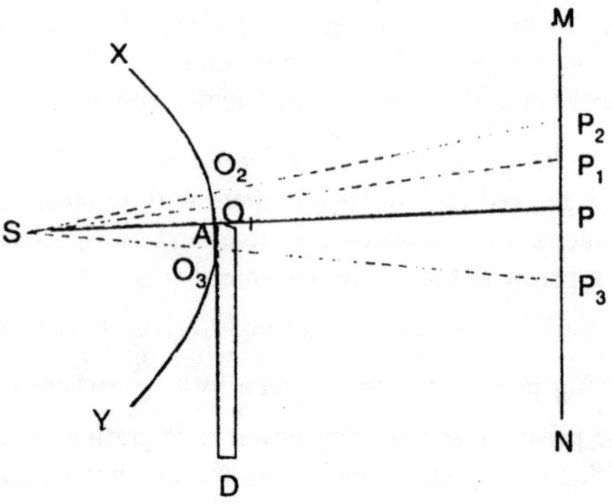

Fig. 3.48

Let us consider the points P, P_1 and P_2 in the illuminated portion and point P_3 in the geometrical shadow region of the screen MN. A, O_1, O_2 and O_3 are the poles of the wave front for the points P, P_1, P_2 and P_3 respectively.

Intensity at the Point P_1

For the point P_1, O_1 is the pole of the wave front and let the exposed portion of the wave front between A and O_1 correspond to the spiral OM_1 (Fig. 3.49). The intensity in this case is proportional to the square of the vector M_1J_2. Thus, the point P_1 will be of maximum intensity.

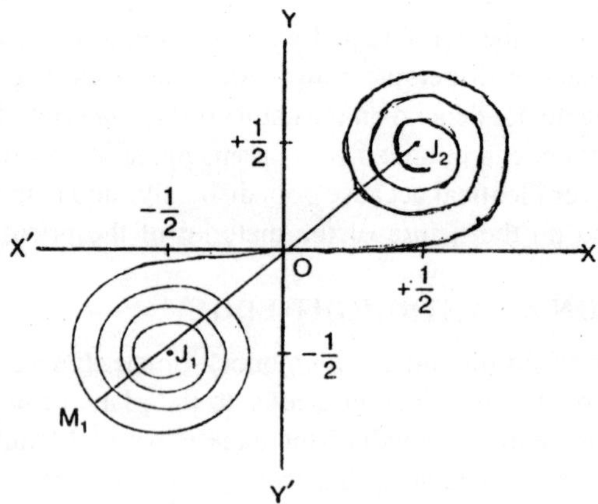

Fig. 3.49

Intensity at the Points P_2

For the point P_2, O_2 is the pole of the wave front and let the exposed portion of the wave front between A and O_2 correspond to the length of the spiral OM_2 (Fig. 3.50). The intensity at P_2 is proportional to the square of the amplitude vector M_2J_2 which is a minimum.

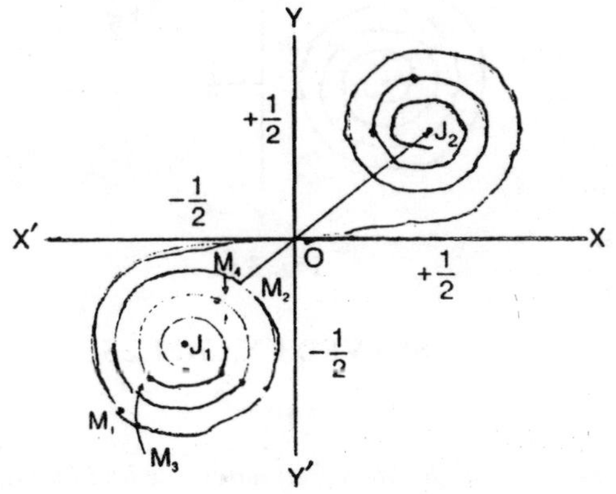

Fig. 3.50

Thus, when the point shifts away from P, the resultant intensity is proportional to the square of the amplitude vector whose magnitude passes through maxima and minima.

When the point M shifts very near J_1, the difference in intensity between maxima and minima is very small and this results in uniform illumination.

Intensity at the Point P_3

For the point P_3, 0_3 pole of the wave front. The lower half of the wave front is cut off by the obstacle and the portion of the wave front between A and O_3 is also obstructed corresponding to the length of the spiral OM (Fig. 3.51).

The intensity at P_3 is proportional to the square of the amplitude MJ_3. As the point P_3 shifts more and more into the region of geometrical shadow, the point M shifts more and more towards J_2. Thus, the magnitude of MJ_2 gradually decreases. In other words, the intensity falls off gradually in the region of the geometrical shadow.

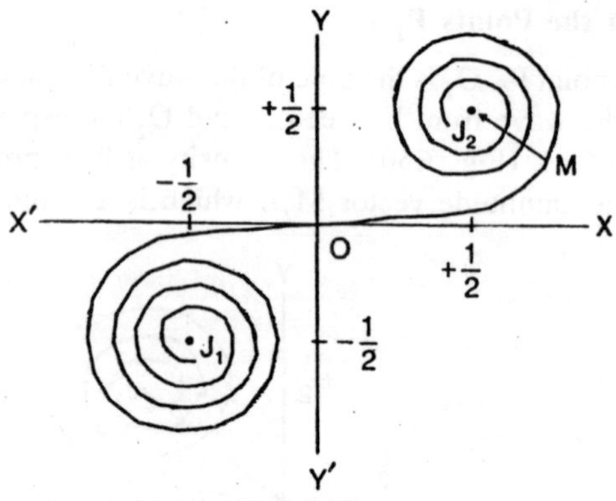

Fig. 3.51

SOLVED EXAMPLES

Example 1:

Microwaves of 6000 Mc/s frequency are incident normally on a slit of variable width and large length. A receiver is placed at a large distance on the other side at angle 20° with the normal. For what widths of the slit would the receiver show zero intensity?

Solution:

Speed of microwaves = speed of light = 3×10^{10} cm/sec

Hence $\lambda = \frac{c}{v} = \frac{3\times10^{10}}{6\times10^{9}} = 5.0$ cm

Let the variable slit width be a. Then for minima

$$a \sin 20° = n \times 5.0; \; (n = \pm 1, \pm 2...)$$

sin 20° = 0.34. Substitution gives

$$a = \frac{5n}{0.34} = 15, 29, 44 \; ... \text{ cm.}$$

Example 2:

Two parallel slits have widths 10 mm each and opaque space between them is 20 mm. If light of wavelength 5 m (this is in microwave region) falls normally on the system, deduce (i) the angular spread of the central diffraction maximum, (ii) the angular positions of the first 4 interference maxima on one side.

Solution:

a = 1.0 cm. e = 3.0 cm. λ = 0.5 cm.

Spread of central diffraction maximum is given by

$$\theta = \pm \sin^{-1}(\lambda/a) = \pm 30°$$

The interference maxima are given by

$$\theta = \sin^{-1}(n\lambda/e) = \sin^{-1}(n/6)$$

Apart from the zero order maximum (θ = 0), the four maxima on one side occur at

$$\theta = \sin^{-1}(1/6) \quad \sin^{-1}(2/6) \quad \sin^{-1}(3/6), \quad \sin^{-1}(4/6)$$

$$= 9°36', \quad 19°30', 30°0, \quad 41°48',.$$

However, the interference 'maximum' due to appear at 30° has zero intensity since the first 'zero' of diffraction falls there.

Example 3:

A plane transmission grating has e = 2 × 10^{-4} cm. Deduce the positions of first, second and third order maxima for:

(i) λ = 5 × 10^{-5} cm,

(ii) λ = 8 × 10^{-5} cm. Assume normal incidence.

Solution:

We have

(i) For $\lambda = 5 \times 10^{-5}$ cm, $\sin\beta = \frac{n\lambda}{e}$ = 0.25, 0.50, 0.75 (check)

$\therefore \quad \beta$ = 14°30', 30°, 48°36'

(ii) For λ 8 × 10^{-5} cm, $\sin\beta = \frac{n\lambda}{e}$ = 0.40, 0.80, (1.20)

$\therefore \quad \beta$ = 23°36', 53°6' (No third order).

Example 4:

A plane diffraction grating has e = 15 × 10^{-4} cm. Calculate the position of the third order maximum for λ = 2.4 × 10^{-4} cm. (This wavelength is in infra-red region). For which other longer wavelengths do the maxima fall in the same position? Assume normal incidence.

Solution:

$$\sin\beta = \frac{n\lambda}{e} = \frac{3\times 2.4\times 10^{-4}\,\text{cm}}{1.5\times 10^{-4}\,\text{cm}} = 0.48 \Rightarrow \beta = 29°$$

For *longer* wavelengths giving maxima at the same angle, we have to consider smaller n, that is n = 2 and n = 1. If the wavelengths are λ' and λ'' we have

$$3\lambda = 2\lambda' = 1\lambda''$$

$$\Rightarrow \quad \lambda' = 3.6 \times 10^{-4} \text{ cm},$$

$$\lambda'' = 7.2 \times 10^{-4} \text{ cm}.$$

Example 5:

A plane diffraction grating of e = 17 × 10⁻⁵ cm is used to photograph a spectrum using normal incidence. Calculate angular dispersion in the region of the different orders of the spectrum of λ = 5.0 × 10⁻⁵ cm. If the camera has focal length 25 cm, calculate the linear dispersion in the spectrograph, and also the separation between spectral lines 5890 Å and 5896 Å in the second order.

Solution:

For the three orders–

$$\beta = \sin^{-1}\left(\frac{n\lambda}{e}\right) = \sin^{-1}(.311), \sin^{-1}(.622), \sin^{-1}(.933)$$

$$= 18°6', 38°30', 68°54',$$

Corresponding values of cos β are 0.951, 0.783, 0.360

$$\frac{\Delta\beta}{\Delta\lambda} = \frac{n}{e\cos\beta} = 5.7 \times 10^3, 13.4 \times 103, 43.9 \times 10^3 \text{ rad/cm}.$$

$$= 5.7 \times 10^{-5}, 13.4 \times 10^{-5}, 43.9 \times 10^{-5} \text{ rad/Å}.$$

Linear dispersion is f. Δβ/Δλ. Calling it Δx/Δλ,

$$\frac{\Delta x}{\Delta\lambda} \text{ for second order} = 25 \times 13.4 \times 10^{-5} \text{ cm/Å}$$

For Δλ = 6A° we get

$$\Delta x = \frac{\Delta x}{\Delta\lambda}.\ \Delta\lambda = 6 \times 3.35 \times 10^{-3} = 2.04 \times 10^{-2} \text{ cm}.$$

Example 6:

In the photograph of a spectrum three lines are observed at positions 12.3472, 12.3894 and 12.4037 cm. The first and last are known to be second order spectra of λ = 5270.81A° and 5272.14A° respectively. Deduce :

(i) The linear dispersion, and

(ii) The possible wavelength for the spectral line at 12.3894 cm.

Solution:

$$\text{Linear dispersion} = \frac{\Delta x}{\Delta\lambda} = \frac{0.0565\text{cm}}{1.33\text{A}^\circ} = 0.04325 \text{ cm per A}^\circ$$

Let the third line also be the *second* order spectrum of some λ. For small regions of λ we can use linear dispersion. Hence

$$\lambda = 5270.81\text{A}^\circ + \frac{(12.3894 - 12.3472)\text{cm}}{0.0424\,\text{cm}/\text{A}^\circ} = 5271.80\text{A}^\circ$$

But this line could also be the first order spectrum of 2 × 5271.80, or third order spectrum of 2/3 × 5271.80Å etc. Thus, the *possible* λ values are

10543.60Å, 5271.80Å, 3541.53Å...

Example 7:

Calculate the possible order of spectra with a plane transmission grating having 18,000 lines per inch when light of wavelength 4500Å is used.

Solution:

Order of spectra, $n = \frac{d\sin\theta}{\lambda}$. The highest order occurs when $\sin\theta = 1$;

$$\therefore\ n = \frac{d}{\lambda} = \frac{1}{N\lambda} = \frac{1}{(7.09\times10^5\,\text{lines}/\text{m})(4500\times10^{-10}\,\text{m})} = 3$$

Example 8:

Light which is a mixture of two wavelengths 5000Å and 5200Å is incident normally on a plane transmission grating having 10000 lines per cm. A lens of focal length 150 cm is used to observe the spectrum on the screen. Calculate the separation in cm of the two lines in the first order spectrum.

Solution:

Here $(a + b) = 10^{-4}$ cm, $l_1 = 5000Å = 5 \times 10^{-5}$ cm;

$\lambda_2 = 5200Å = 5.2 \times 10^{-5}$ cm and $n = 1$

$$\text{Now} \quad \sin\theta_1 = \frac{n\lambda_1}{a+b} = \frac{1\times5\times10^{-5}\,\text{cm}}{10^{-4}\,\text{cm}}$$

$$\therefore\ \theta_1 = \sin^{-1}(0.5) = 30^\circ$$

Similarly, $\sin\theta_2 \frac{\cdot n\lambda_2}{a+b} = \frac{1\times 5.2\times 10^{-5}\,cm}{10^{-4}\,cm}$

$= 0.52$

$\theta_2 = \sin^{-1}(0.52) = 31.3°$

Further, $\tan\theta_1 = \frac{x_1}{f}$ and $\tan\theta_2 = \frac{x_2}{f}$

$\therefore\ (x_2 - x_1) = f\,[\tan\theta_2 - \tan\theta_1]$

$= 150\,[0.0687 - 0.5774]$

$= 4.7$ cm.

Example 9:

A zone plate is to be constructed with focal length 50 cm for $\lambda = 5.0 \times 10^{5}$ cm. Calculate its first radius. What will be its focal length for $\lambda = 4.0 \times 10^{5}$ cm?

Solution:

$r_1^2 = f\lambda = 50 \text{ cm} \times 5.0 \times 10^{-5} \text{ cm}$

$\therefore\ r_1 = 5.0 \times 10^{-2} \text{ cm} = 0.50 \text{ mm}$

$f_2 = f_1, \frac{\lambda_1}{\lambda_2} = 50 \times \frac{5}{4} = 62.5$ cm.

Example 10:

In an experiment with straight edge diffraction, the slit to edge distance is 1.00 metre, and edge to screen distance is 2.00 metres. If l = 5000 Å, deduce the separation of the first three bright fringes.

Solution:

The *positions* of the maxima (relative to geometrical projection of the edge) are given by

$x_n = K\sqrt{n}$, where n = 1, 3, 5...

and $k = \left[\frac{b(a+b)\lambda}{a}\right]^{\frac{1}{2}}$

We have $a = 100$ cm, $b = 200$ cm, $\lambda = 5 \times 10^{-5}$ cm

$\therefore\ K = \left(\frac{200\times 300\times 510^{-5}}{100}\right)^{\frac{1}{2}} = 0.171$ cm

The first three maxima therefore lie at

$$0.171,\ 0.171\sqrt{3},\ \text{and}\ 0.171\sqrt{5}\ \text{cm}$$

The *separations* are

$$0.171\sqrt{3} - 0.171 = 0.125\ \text{cm}$$

$$0.171\sqrt{5} - 0.171\sqrt{3} = 0.086\ \text{cm}.$$

Example 11:

A circular aperture of 1.2 mm diameter is illuminated by plane waves of monochromatic light. The diffracted light is received on a distant screen which is gradually moved towards the aperture. The centre of the circular patch of light first becomes dark when the screen is 30 cm from the aperture. Calculate the wavelength of light.

Solution:

Given that diameter = 1.2 mm = 0.12 cm

Radius r = 0.06 cm

$$b = 30\ \text{cm}$$

Here $$(b^2 + r^2) = (b + l)^2$$

$$302 + (0.06)^2 = (30 + l)^2$$

$$\lambda = \frac{(0.06)^2\ \text{cm}^2}{2 \times 30\text{cm}} (\lambda^2 \text{neglected})$$

$$= 0.00006\ \text{cm} = 6000\text{Å}.$$

Example 12:

A plane wave front of light of wavelength 5.5 × 10 ⁵ cm is intercepted by a circular aperture of 5.0 mm diameter. Calculate the number of half-period zones in the aperture for an axial observation point distant

(i) 110 metre,

(ii) 5.0 metres from the aperture.

Solution:

For case (i),

$$m = \frac{(0.25\text{cm})^2}{(100\text{cm})(5.5 \times 10^{-5}\text{cm})}$$

= 11.4 h.p. zones

For case (ii) be is 5 times larger. Hence

$$m = \frac{11.4}{5} = 2.3 \text{ h.p. zones.}$$

Example 13:

What is the radius of the first half period zone in a zone plate behaving like a convex lens of focal length 60 cm for light of wavelength 6000Å?

Solution:

Given that f = 60 cm = 0.6 m, n = 1,

and λ = 6000Å = 6×10^{-7}m

$$f = \frac{r_n^2}{n\lambda} \quad \therefore r_n^2 = fn\lambda$$

$$\therefore \quad r_1^2 = 0.6 \text{ m} \times 1 \times 6 \times 10^{-7}\text{m}$$

$$= 36 \times 10^{-8}\text{m}^2$$

$$\text{or} \quad r_1 = 6 \times 10^{-4}\text{m} = 0.6 \text{ mm.}$$

Example 14:

Diffraction pattern of a single slit of width 0.5 cm is formed by a lens of focal length 40 cm. Calculate the distance between the first dark and the next bright fringe from the axis. Wavelength of light used is 4890Å.

Solution:

Here, a = 0.5 cm = 5×10^{-3}m, f = 40 cm = 0.4 m,

λ = 4890Å = 4890×10^{-10}m

For a minima, a sin θ = nλ.

Also $\sin\theta = \frac{x_1}{f}$

As n = 1, we get $\frac{x_1}{f} = \frac{\lambda}{a}$

$$\text{or} \quad x_1 = \frac{f\lambda}{a} = \frac{0.4 \times 4890 \times 10^{-10}}{5 \times 10^{-3}} \text{m}$$

$$\therefore \quad x_1 = 3.912 \times 10^{-5} \text{ m.}$$

Now for a maximum, $a \sin\theta = \frac{(2n+1)\lambda}{2}$.

Also $\sin\theta = \frac{x_2}{f}$

As n = 1, we get $\frac{x_2}{f} = \frac{3\lambda}{2a}$

or $x_2 = \frac{3\lambda f}{2a}$

$\therefore \quad x_2 = \frac{3 \times 4890 \times 10^{-10} \times 0.4}{2 \times 5 \times 10^{-3}} \text{m} = 5.868 \times 10^{-5} \text{ m}$

$\therefore \quad x_2 - x_1 = 5.868 \times 10^{-5} \text{ m} - 3.912 \times 10^{-5} \text{ m}$

$= 1.956 \times 10^{-5} \text{ m} = 1.9 \times 10^{-2} \text{ mm.}$

Example 15:

Deduce the missing orders for a double slit Fraunhofer diffraction pattern, if the slit widths are 0.16 mm and they are 0.8 mm apart.

Solution:

Here a = 0.16 mm = 0.016 cm and b = 0.8 mm = 0.08 cm

Equation for interference maxima is, $(a + b) \sin\theta = n\,\theta$

Equation for diffraction minima is, $a \sin\theta = p_1$

$\therefore \quad \frac{(a+b)}{a} = \frac{n}{p}$

$\therefore \quad \frac{n}{p} = \frac{(0.016 + 0.080)}{0.016} = 6$

$\therefore \quad n = 6p$

For the values of p = 1, 2, 3 etc.

n = 6, 12, 18 etc.

Thus the orders 6, 12, 18 etc. the interference maxima will be missing from the diffraction pattern.

Example 16:

In a diffraction phenomenon using double slit, calculate :

(i) The distance between the central maximum and the first minimum of the fringe envelope, and

(ii) The distance between any two consecutive double slit dark fringes.

Given data: Wavelength of light = 5000Å,

Slit width = 0.02 mm,

Spacing between two slits = 0.10 mm,

Screen to slits distance = 100cm

Solution:

Here, a = 0.02 mm = 2×10^{-5}m,

b = 0.1 mm = 10^{-2} m, (a + b) = 1.2×10^{-4} m,

l = 5000Å, d = 100 cm = λm

(i) The angular separation between the central maximum and the first minimum is,

$$\sin\theta_1 = \theta_1 = \frac{\lambda}{2(a+b)} \text{ and } \theta_1 = \frac{x_1}{D}$$

$$\therefore \quad \frac{x_1}{D} = \frac{\lambda}{2(a+b)} \quad \therefore x_1 = \frac{\lambda}{2(a+b)}$$

$$\therefore \quad x_1 = \frac{5\times10^{-7}\,m\times1m}{2(1.2\times10^{-4}\,m)} = 2.08 \times 10^{-3}\ m = 2.08\ mm$$

The separation between the central maximum and the first minimum is 2.08 mm.

(ii) The angular separation between two consecutive dark fringes.

$$\sin\theta_1 - \sin\theta_2 = \theta_1 - \theta_2 = \theta = \frac{3\lambda}{2(a+b)} - \frac{\lambda}{2(a+b)}$$

$$\theta = \frac{\lambda}{(a+b)}; \qquad \text{Also } \theta = \frac{x_2}{D}$$

$$\therefore \quad x_2 = \frac{\lambda D}{(a+b)} \qquad \therefore x_2 = \frac{5\times10^{-7}\,m\times1m}{1.2\times10^{-4}\,m}$$

= 4.16×10^{-3} m = 4.2 mm

Example 17:

A parallel beam of wavelength 5460Å is incident at an angle of 30° on a plane transmission grating which has 6000 lines/cm. Find the highest order spectrum that can be observed.

Solution:

Here $\theta = 30°$, $\lambda = 5460 Å = 5460 \times 10^{-10}$m,

$$(a + b) = \frac{1}{6\times10^3}\text{cm} = \frac{1}{6\times10^5}\text{m}$$

Now, $(a + b)[\sin\theta_n + \sin i] = n\lambda$

But here $\theta_n = i$

$\therefore$ $(a + b)[2\sin i] = n\lambda$

$$\therefore \quad n = \frac{(a+b)[2\sin i]}{\lambda} = \frac{1}{6\times10^5 5460\times10^{-10}} = 3.05$$

or $n = 3$.

Example 18:

Sound waves of frequency 3000 Hz fall normally on an opening of width 0.50 metre. Calculate the angular positions of the diffraction minima. (Take speed of sound as 345 m/sec).

Solution:

$$\lambda = \frac{345}{3000} = 0.115 \text{ m}, \quad \frac{\lambda}{a} = 0.23$$

Minima occur at $\theta = \pm \sin^{-1} n\lambda/a$. Since maximum value of $\sin\theta$ can be 1, the only minima are at,

$\pm \sin^{-1} 0.23$, $\pm \sin^{-1} 0.46$, $\pm \sin^{-1} 0.69$, $\pm \sin^{-1} 0.92$

or $\pm 13°$, $\pm 27°$, $\pm 44°$, $\pm 67°$

Note if an observer walks across at a large distance D from the opening, then starting from a point on the perpendicular bisector of the opening he will observe minima of sound at *positions* given by D tan θ. If for example, D = 4 metre, the minima appear at

$\pm$ 0.92m, $\pm$ 2.0m, $\pm$ 3.8m, $\pm$ 9.9m.

Example 19:

A set of 10 parallel equidistant slits of width 0.50 cm each and spacing e = 1.90 cm are used to study Fraunhofer diffraction of waves of λ = 0.60 cm falling normally on the plane of the slits. Deduce :

(i) angular position of the interference maxima,

(ii) half-widths of these maxima,

(iii) the effect of covering up 5 slits from one end,

(iv) the effect of covering up alternate slits.

Solution:

The data given are:

$$a = 0.50 \text{ cm}, \quad e = 1.90 \text{ cm},$$

$$\lambda = 0.60 \text{ cm}, \quad N = 10$$

(i) The interference maxima:

$$\sin\theta = \frac{0.60}{1.90}, \frac{2\times 0.60}{.190}, \frac{3\times 0.60}{1.90}$$

$$\therefore \quad \theta = 18°24', \quad 30°10', \quad 71°30'$$

(ii) The half-width

$$\delta\theta = \frac{0.60}{10\times 1.90\cos\theta}$$

$$= .033, .041, .100 \text{ radian.}$$

(iii) This step reduces N to 5, keeping e unchanged. Hence q remains unaltered, dq becomes doubled.

(iv) The step reduces N to 5, and doubles e. Hence the q deduced above now correspond to II, IV and VI order maxima, their half widths remain unaltered. (Positions and half-widths of I, III, V orders may be deduced too).

Example 20:

A concave grating of 6000 lines/cm and radius of curvature 5.0 metre is used to photograph sodium D – lines (5890, 5896 A) in the second order with Rowland mounting.; Calculate the linear separation of the doublet on the photograph.

Solution:

The angular dispersion is given by

$$\frac{\delta\beta}{\delta\lambda} = \frac{n}{e\cos\beta} = \frac{n}{e} \text{ (since } \beta = 0 \text{ in Rowland mounting)}$$

Linear separation $\delta x = R\delta\beta$. Hence from the given data

$n = 2$, $e = (1/6000)$ cm,

$\delta\lambda = 6 \times 10^{-8}$ cm and $R = 500$ cm.

Example 21:

Along the axis of a circular aperture of diameter 2.00 mm on one side is a point source at 1.00 metre distance emitting light of $\lambda = 5.46 \times 10^{-5}$ *cm, and on the other side is a point of observation at a varying distance D. For what valued of D does the aperture involve (i) two, (ii) four, (iii) six half-period zones?*

Solution:

We are given r = 0.100 cm, a = 100 cm,

$\lambda = 5.46 \times 10^{-5}$ cm

and b = D, which varies as n varies.

$$\therefore \frac{1}{100} + \frac{1}{D} = \frac{1}{100} + \frac{1}{D} = m \times 5.46 \times 10^{-3} \text{ cm}^{-1}$$

For m = 2, 4, 6,

we get D = 1.09×10^3, 84.4, 43.8 cm respectively (check).

Example 22:

In Fraunhoffer diffraction pattern due to a narrow slit a screen is placed 2m away from the lens to obtain the pattern. If the slit width is 0.2 mm and the first minima lie 5 mm on either sides of the central maximum, find the wavelength of light.

Solution:

Here, a = 0.2 mm = 0.02 cm,

x = 5 mm = 0.5 cm,

D = 2 m = 200 cm

In the case of Fraunhofer diffraction at a narrow rectangular aperture,

$a \sin\theta = n\lambda$

Here n = λ $\therefore$ $a \sin\theta = \lambda$

$$\sin\theta = \frac{x}{D} \therefore \lambda = \frac{ax}{D}$$

$$\text{or} \qquad \lambda = \frac{0.02\text{cm} \times 0.5\text{cm}}{200\text{cm}}$$

$$= 5 \times 10^{-5} \text{ cm} = 5000\text{Å}$$

EXERCISES

1. Discuss the properties of Cornu's spiral and explain its relationship with Fresnel's half period zones. Show how the spiral can be used to obtain the intensity distribution in the Presnel diffraction pattern due to a straight edge.

2. The diameter of the central zone of a zone plate is 3 mm. If a point source of light (λ = 6000 Å) is placed at a distance of 5m from it, calculate the position of the first image.

3. A parallel beam of monochromatic light of λ, = 5000Å is incident normally on a plate having a circular hole of diameter 1mm. The screen is at the farthest position for which the axial point is almost black. The screen is moved towards the plate so that the axial point is again seen black. How far is the screen moved from the first position to the second?

4. The phasor diagram forms a spiral shape in one case and circular shape in the other.

5. A plane diffraction grating has 5000 lines per cm. How many orders of spectra will it show for $\lambda = 5 \times 10^{-5}$ cm using :

   (i) normal incidence,

   (ii) normal observation,

   (iii) minimum deviation?

   Also deduce which wavelengths in the range 3500 to 8000.4 will overlap with the third order for $\lambda = 5 \times 10^{5}$cm.

6. A narrow slit is illuminated with light of $\lambda = 5.89 \times 10^{-5}$ cm. A straight edge is placed 2.0 m away, the edge being parallel to the slit. Observations are made 3.0 m from the edge. Compute the positions of the first three maxima and the intervening minima relative to the 'geometrical edge' of the shadow.

7. Figure below is a spiral and it gives (not to scale) the phasor diagram for a spherical wavefront. Explain, why this differs so much with the phasor diagram for a cylindrical wavefront (Cornu's spiral).

8. The diameter of the first ring of a zone plate is 1.1 mm. If plane waves (λ = 6000Å) fall on the plate, where should the screen be placed so that light is focused to a bright spot?

9. Find the radii of the first three transparent zones of a zone plate whose first focal length is 1 m for λ, = 5893Å.
10. Explain the difference between Fraunhoffer and Fresnel diffraction.
11. Describe two simple observations which you can make in everyday life to show that light is diffracted.
12. Consider a wavefront divided into one-fourth period zones. If amplitude contributed at the observation point is independent of the zone number n, find by {he phasor method the resultant amplitude due to the first :

    (i) 2 zones,

    (ii) 4 zones,

    (iii) 6 zones,

    (iv) 8 zones. Repeat, taking amplitude varying as $n^{-1/2}$.
13. Describe and explain the phenomenon of diffraction due to a straight edge. Explain why the bands are neither equidistant nor equally illuminated.
14. A sharp razor blade is held vertically in a beam of monochromatic light coming from a narrow slit parallel to the edge of the blade. Discuss the position and intensities observed on a screen placed behind the blade. How will you use it to determine the wavelength of light?
15. I_θ is proportional to N^2. But doubling the number of slits would only double the available energy. How do you reconcile that I_θ is not proportional to N but to N^2?
16. Consider a circular aperture of radius 1.0 cm. For a point source of λ = 5000Å at infinite distance along the axis, deduce the positions on the axis on the other side where the intensity would be zero. Why would this be difficult to observe? Repeat the calculation with aperture radius = 0.1 cm.
17. An object placed 50 cm away from a zone plate gives a real image at 70 cm on the other side for λ = 5000Å Calculate :

    (i) position of the image for λ = 7000Å,

    (ii) positions of the next two secondary images for λ + 5000Å,

    (iii) approximate intensities of secondary images relative to the primary.

18. Deduce the principal focal length of a zone plate whose central zone has radius 0.62 mm and light used has λ = 5000Å. Deduce also the outer radius of 9th h.p. zone.
19. A point source of λ = 6000 Å is placed 50.0 cm away along the axis of a circular aperture 4.0 mm in dia. Locate the axial position on the other side so that the aperture covers 16 h.p. zones. Calculate also the diameters of the first and second bright rings around the central spot there.
20. If all parts of a given h.p. zone are assumed to send waves in same phase, the amplitude contribution is .R. Show that the actual contribution will be $2R/\pi$ if the phase variation for waves from within a h.p. zone is taken into account. State the assumptions involved j in such derivation.
21. Wire-gratings are used in far infra-red. In one case 1000 wires of diameter 0.2 mm each are stretched parallel to one another leaving spacings of 0.05 mm each. Deduce the wavelength for which the third- order maximum falls at 30°, incident light falling normally.
22. A spectrum is seen with a plane diffraction grating. How will this be affected if :
    (a) the grating is moved in its own plane,
    (b) the grating is turned in its own plane,
    (c) the grating is moved perpendicular to its own plane,
    (d) the grating is turned about an axis parallel to the lines on it?
23. A blazed grating has 6000 lines per cm and is blazed for the third order of λ = 25000A°. Calculate the blaze angle.
24. In a cylindrical wavefront the successive equal-phase zones have areas proportional to $\sqrt{m+1} - \sqrt{m}$, where m = 0, 1, 2,... Make a list of successive areas upto m =8. Now consider one-fourth period zones, and deduce the resultant relative intensity due to 1, 2,3,... 8 such zones (may use phasor diagram method).
26. Derive an expression for the intensity at a point in the Fraunhoffer type of diffraction produced by two nearby parallel narrow slits. Draw a diagram to indicate the intensity distribution in this case.
27. In a diffraction grating how are the spectral lines affected when the rulings are made closer.

28. Give the theory of a concave reflection grating. Deduce the condition to obtain focused spectra. Explain why a concave grating is preferred over a plane grating.
29. A plane diffraction grating may be set for viewing a spectrum 'minimum deviation' condition. If the minimum deviation is ϕ, that 2e sin $1/2\phi = n\lambda$. What may be the advantage in minimum deviation setting?
30. What do you understand by diffraction of waves? Explain why the diffraction of sound is readily observed than the diffraction of light.
31. Explain why gratings with larger number of lines are preferred?
32. In a concave grating with Rowland mounting the spectrum shows four lines whose positions on the photographic plate are 0.2174 mm, 0.1493 mm, 0.0897 mm, and 0.0127 mm. If the wavelengths of the first and last lines among the four are 5134.6 A^{0} and 5172.3 Å, calculate the wavelengths of the other two lines.
33. A conçave .grating with 6000 lines per cm and radius of curvature 2.4 meters is used to study the spectrum near 4000 Å in the second order. Deduce the linear dispersion if the mounting used is Rowland's.
34. "The theory of plane diffraction grating does not depend on assuming that widths a are fully transmitting and widths b are fully opaque. It only requires that there are variations in the transmitting power with repetitive spacing e. Comment on this.
35. In what respect is an echelon grating superior to an ordinary ruled grating? Obtain expressions for these in the case of a plane transmission grating.
36. In what respect is an echelon grating superior to an ordinary ruled grating? Give an account of its theory and its practical applications.
37. Derive an expression for the angular dispersion of a plane diffraction grating.
38. Distinguish between Fresnel and the Fraunhofer classes of diffraction. Compare the diffraction phenomenon in the following cases:

    (a) Diffraction by a slit and a narrow wire,

(b) Diffraction by a circular hole and a circular obstacle.

39. Give the theory of a diffraction grating. Describe in detail how you would use a transmission grating for measuring the wavelength of light.
40. Discuss the diffraction of light by a narrow slit.
41. Discuss Fraunhofer diffraction due to a single slit. Extend the theory to the case of a plane transmission grating. Explain what is meant by diffraction spectra of different orders and state the conditions under which the grating spectra of even order are absent.
42. Discuss Fraunhofer diffraction pattern of a straight edge. How does this pattern differ from that due to a straight wire?
43. Give an account of the diffraction effects produced by a slit. Explain what happens when the slit width is gradually increased and also when the screen is gradually moved away from the slit.
44. Explain the formation of spectra by a plane diffraction grating. What are its chief characteristics?
45. Discuss Fraunhofer diffraction due to a single slit. Draw a curve indicating distribution of light in the diffraction pattern. Is there any fundamental difference between interference and diffraction? Give reasons.
46. Describe and explain the phenomenon of diffraction due to a straight edge. Explain why the bands are neither equidistant nor equally illuminated.
47. Explain how plane transmission grating spectra can be used to determine the wavelength of sodium light. Compare the grating spectra with prism spectra.
48. Explain the formation of spectra by a plane diffraction grating. What particular spectra would be absent if the width of the transparencies and opacities of the grating are equal?
49. Give the theory of the formation of the spectra of the various orders on the Rowland's circle by a concave reflection grating.
50. Explain why a grating is designed to produce only two orders. Why the grating spectrum is called a normal spectrum?
51. Discuss the theory of a plane diffraction grating for oblique incidence.

52. In a Fraunhofer diffraction due to a single slit, show that the intensity of the first secondary maximum is roughly 4.5% of that of the principal maximum.

53. Explain giving suitable examples, what is meant by Fraunhofer class of diffraction. Describe in detail how you would use this phenomenon for determining the wavelength of light.

54. Find the half angular width of the central bright maximum in the Fraunhoffer diffraction pattern of a slit of width 12×10^{-5}cm when the slit is illuminated by monochromatic light of wavelength 6000Å.

55. In a single slit diffraction pattern the distance between the first minima on either side of die central zero maximum is 4.4 mm as observed on a screen at a distance of 0.7 m. The wavelength of light used is 5890Å. Calculate the slit width.

56. A single slit illuminated by red light of 6000Å wavelength gives first order Fraunhoffer diffraction minima that subtends angle of 4° with the axis. How wide is the slit?

57. Light of wavelength 6000Å is incident on a slit of width 0.30 mm. The screen is placed 2 m away from the slit. Find (a) the position of the first dark fringe and (b) the width of the central bright fringe.

58. Light of wavelength 6000Å is incident on a narrow slit. The screen is placed 2 m away from the slit. If the first minima lie 5 mm on either side of the central maximum, calculate the slit width.

59. In a Fraunhoffer diffraction pattern of a single slit, the screen is placed 100 cm away from the slit and the slit is illuminated by monochromatic light of wavelength 5893Å. The width of slit is 0.1 mm. Calculate the separation between the central maximum and the first secondary minimum.

60. In a Fraunhoffer diffraction phenomenon using double slit, calculate :

(i) the fringe spacing on a screen, and

(ii) distance between the central maximum and the first minimum of the fringe envelope.

Given data: Screen to slits distance = 50 cm wavelength of light = 4800Å, slit width = 0.02 mm, spacing between two slits = 0.10 mm.

61. A parallel beam of monochromatic light is allowed to be incident normally on a plane transmission grating having 5000 lines/cm and the second order spectral line is found to be diffracted through 30°. Calculate the wavelength of light.

62. A diffraction grating used at normal incidence gives a line λ_1 = 6000Å in a certain order superimposed on another line λ_2 = 4500Å of the next higher order. If the angle of diffraction is 30°, how many lines are there in a cm in the grating?

63. A parallel beam of sodium light (λ = 5890Å) is allowed to be incident normally on transmission grating and second order spectra lines are found to be derived through 30°. Calculate the number of lines per cm on the grating.

64. Describe and explain the Fraunhofer pattern obtained with a narrow slit and illuminated by a parallel beam of monochromatic light.

65. Explain the construction and the mode of action of a diffraction grating and derive an expression for its resolving power.

66. Give the theory of a concave grating. Give a short account of the important methods of mounting the gratings with their respective advantages and disadvantages.

67. A parallel beam of monochromatic light is allowed to be incident normally on a plane transmission grating having 15000 lines/inch. Find the angle of separation of the 5048Å and 5016 A lines of helium in the second order spectrum.

68. A Rowland grating has 6000 lines per cm. Calculate the angular separation of the two mercury lines of wavelengths 5770Å and 5461Å in the first order spectrum.

69. Deduce the area of a h.p. zone for a = 30 cm, b = 150 cm, $\lambda = 6 \times 10^{-5}$ cm in case of spherical wavefronts. Deduce the two smallest radii of a circular aperture for which the observation point will be (i) dark, (ii) bright.

4

Propagation of Light Waves

INTRODUCTION

The physical laws describing propagation, reflection, refraction, attenuation of electromagnetic waves accurately describe the behaviour of light waves. In the electromagnetic theory of light, the mechanical displacement of the medium is replaced by a variation of the electric field at the corresponding point. The light wave consists of varying electric and magnetic fields and can be described by two vectors, the amplitude of electric field strength, E, and the amplitude of the magnetic field strength, H. These vectors oscillate at right angles to each other and to the direction of propagation. They cannot be separated. Thus, a beam of light is a traveling configuration of electric and magnetic fields. The electromagnetic theory was developed by James Clerk Maxwell in 1873. The first and foremost outcome of Maxwell's equations was the prediction of existence of electromagnetic waves and the brilliant prediction that light waves are electromagnetic waves. Starting from the Maxwell's equations, the propagation of light waves in space and materials can be easily explained and all the fundamental laws of optics such as law of reflection, Snell's law of refraction etc. can be derived. Expressions to calculate the reflectivity, transmissivity and absorptivity of material media can be obtained from these equations. Further, the electromagnetic theory explains the origin of refractive index and the dispersion properties of optical media.

Electromagnetic Waves

Maxwell showed that by combining the four equations a *wave equation* was obtained which described the propagation of waves. The time variation of a magnetic field induces an electric field, while a variation of an electric field, in its turn, induces a magnetic field and electromagnetic fields can exist independently, without electric charges and currents. The continuous inter-conversion of the fields preserves them and an electromagnetic perturbation propagates in space. Such fields are called *electromagnetic waves*. The generation of electromagnetic wave does not require any medium. *The electromagnetic waves propagate through space entirely on their own.* Maxwell's theory placed no restriction on possible wavelengths for the electromagnetic radiation. The vector cross products in Maxwell's second and fourth equations imply that the two fields, E and H are normal to each other and also normal to the direction of propagation. In a vacuum, these waves always propagate at a velocity equal to the velocity of light c.

The vectors E and B in an electromagnetic wave always oscillate in phase (Fig. 4.1). The instantaneous values of E and B at any point are connected through the relation

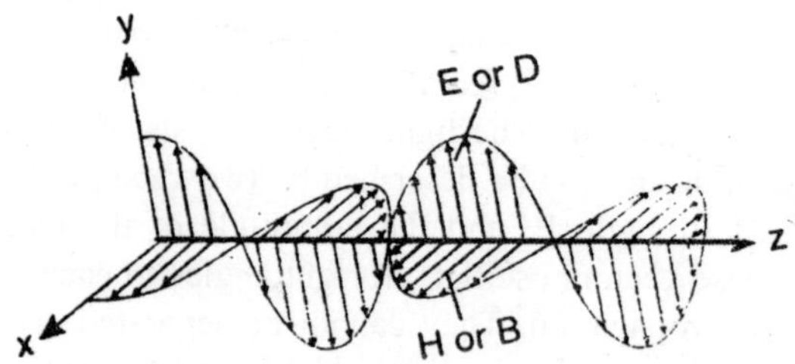

Fig. 4.1 : The directions of E and H in a uniform plane wave.

$$\sqrt{\varepsilon}\, E = \sqrt{\mu}\, H.$$

This means that E and H (or B) simultaneously attain their maximum values, vanish etc.

Note that electromagnetic waves are, in general, neither longitudinal nor transverse. In some types of electromagnetic waves, the electric vector E is at right angles to the ray while the magnetic vector H is not. Waves of this type are called *transverse electric waves* or TE waves. Since H vector is not normal to the ray, it must have a component along the ray direction. In another type of waves the magnetic vector is at right angles to the ray while the electric vector E is not. Waves of this type are called *transverse magnetic waves* or TM waves. If both the vectors E and H are at right angles to the ray, the waves are called *transverse electromagnetic waves* or T_{EM} waves.

Physical Significance of Maxwell's Equations

The physical significance of Maxwell's equations can be readily interpreted from their mathematical statement in the integral form.

(1) Maxwell's first equation (1) shows that the total electric flux density D through the surface enclosing a volume is equal to the charge density p within the volume. It means that a charge distribution generates a steady electric field.

(2) The second equation shows that the e.m.f. around a closed path is equal to the time derivative of the magnetic flux density through the surface bounded by the path. It means that an electric field can also be generated by a time-varying magnetic field.

(3) Maxwell's third equation tells us that the net magnetic flux through a closed surface is zero. It implies that magnetic poles do not exist separately in the way as electric charges do. Thus, in other words, magnetic monopoles do not exist.

(4) Maxwell's fourth equation shows that the magneto motive force around a closed path is equal to the conduction current plus the time derivative of the electric flux density D through any surface bounded by the path. The time derivative of the electric flux density $\frac{\partial D}{\partial t}$ is called displacement current. Thus this equation means that a magnetic field is generated by a time-varying electric field.

LIGHT PROPAGATING THROUGH A MEDIUM

When a beam of light is propagated in a material medium its velocity is less than the velocity in a vacuum and its intensity gradually decreases as it progresses through the medium.

The velocity of light in a material medium varies with the wavelength, and this variation is referred to as *dispersion.* As the beam traverses the medium, its intensity decreases gradually.

The reduction in the intensity of a beam traversing a medium is partly due to *scattering* and partly due to *absorption.* In the true absorption the light disappears, the energy being converted into heat. The phenomenon of scattering, absorption and dispersion are inseparably connected.

Absorption

All media show some absorption. Some media absorb all wavelengths more or less equally; they are said to show *general absorption.* Others absorb some wavelengths very much more strongly than others; they are said to show *selective absorption.* If we assume plane waves propagating in the medium, the amplitude and intensity of plane waves will not be constant but will decrease as the wave passes through the medium. For plane waves, the fraction dI/I of the intensity lost in traversing an infinitesimal thickness dx is proportional to dx, so that

$$\frac{dI}{I} = -\alpha\, dx \quad ...(1)$$

where a is a constant and known as the *absorption coefficient.*

The decrease in traversing a finite thickness x, is

$$\int_0^x \frac{dI}{I} = -\alpha \int_0^x dx$$

$$[\ln I]_0^x = -\alpha x + C$$

where C is a constant. If $I = I_o$ at $x = 0$, $C = \ln I_o$ and one has

$$I(x) = I_o e^{-\alpha x} \quad ...(2)$$

This is the exponential *law of absorption* discovered by Bouguer and Lambert independently.

Scattering

Tyndall observed that when a beam of light passes through a transparent substance, solid, liquid or gaseous, a certain portion of light is scattered in all directions.

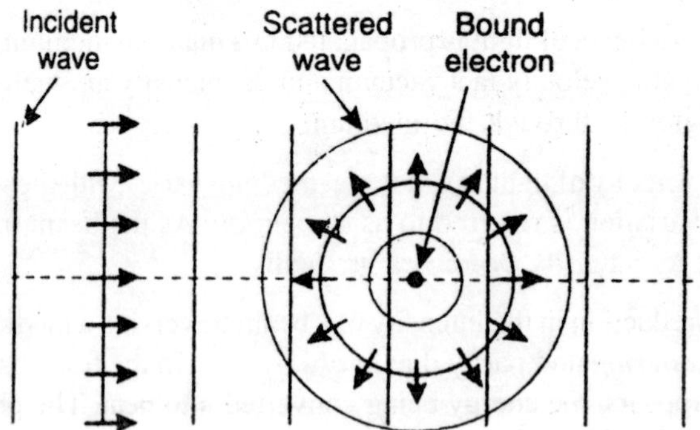

Fig. 4.2 : Scattering of electromagnetic radiation by a bound electron.

This is called the *Tyndall effect.* Tyndall thought-that the faint blue light emerging sideways out of the beam was due to scattering of light by small particles present in the medium. Rayleigh recognized that the scattering was due to individual molecules rather than due to other extraneous particles and such scattering accounts for the blue of the sky. The molecular scattering is hence called the *Rayleigh scattering.* For Rayleigh scattering to take place the molecules or group of molecules should be much smaller in linear dimension than the wavelength of light.

Rayleigh Scattering

A molecule contains an equal number of positive and negative charges. When a molecule is put into a static electric field it suffers some distortion, the positively charged nuclei being attracted towards the negative pole of the field, the electrons to the positive pole. The separation of charge centers causes an *induced electric dipole moment* to be set up in the molecule and the molecule is said to be *polarized.* The size of the induced dipole, μ, depends both on the magnitude of the applied field, E, and on the ease with which the molecule can be distorted. We may write

$$\mu = \alpha E \qquad \text{...(3)}$$

where, a is the polarizability of the molecule. In case of majority of the molecules, the polarizability is anisotropic. That is the electrons forming the bond are more easily displaced by an electric field applied along the bond axis than one across this direction.

When a sample of such molecules is subjected to a beam of light of frequency v the electric field experienced by each molecule varies according to the equation

$$E = E_0 \sin 2\pi vt$$

and thus the induced dipole also undergoes oscillations of frequency v;

$$\mu = \alpha E = \alpha E_0 \sin 2\pi vt \qquad \text{...(4)}$$

Thus, the light filed will make the induced dipole oscillate. Such an oscillating dipole emits light of its own oscillation frequency. In other words, light does not simply pass through transparent matter; instead it forms dipoles, these dipoles radiate and this re-radiation appears as light. The light is seen in any direction perpendicular to the axis of the dipole. This is the same as with a dipole antenna, which radiates in all directions except in the direction of its own length. This is the explanation of *Rayleigh scattering.* The amplitude of the scattered light is inversely proportional to the square of wavelength and since. $I \propto E^2$, the intensity is inversely proportional to the fourth power of the wavelength. Thus,

$$I \propto \frac{1}{\lambda^4} \qquad ...(5)$$

Light scattering is responsible for some of the beautiful colour effects such as blue sky and red sunset.

However, if the incident light is monochromatic, the scattered light is observed to be unchanged in frequency.

Dispersion

The passage of light through a medium causes an electrical disturbance that varies sinusoidally with time. As a consequence, any charged particle in the medium is subjected to a fluctuating electrical field and is set into forced vibrations. Since atoms consist of positive nuclei surrounded by negatively charge clouds of electrons, these differently charged parts of atom will tend to be displaced slightly as a light wave passes. The electrons cannot leave the atom because of the intense local electric field, but they get slightly displaced to one side in a direction opposite to that of the field, as shown in Fig. 4.3a. The nucleus being relatively heavy and thus having a larger inertia will not be moved as much. The electrons being lighter will be forced back and forth in response to the vibrating field. The result is a rapidly oscillating atomic dipole, as shown in Fig. 4.3b.

Thus, when an electromagnetic wave is incident on an atom or a molecule, the periodic electric force of the wave sets the bound charges into vibratory motion. The frequency with which these charges are forced to vibrate is equal to the frequency of the wave. The phase of this motion as compared to the impressed electric force will depend on the impressed frequency. It will vary with the difference between the impressed frequency and the natural frequency of the charges.

Dispersion can be explained with the concept of secondary waves that are produced by the induced oscillations of the bound charges. When a beam of light propagates through a transparent medium (solid or liquid), the amount of lateral scattering is very small. The scattered waves traveling in a lateral direction produce destructive interference.

However, the secondary waves traveling in the same direction as the incident beam superimpose on one another. The resultant vibration will depend on the phase difference between the primary and the secondary waves. This superimposition changes the phase of the primary waves and this is equivalent to a change in the wave velocity. Wave velocity is defined as the speed of which a condition of equal phases is propagated. Hence, the variation in phase due to interference, changes the velocity of the wave through the medium. The phase of the oscillations and hence that of the secondary waves depends upon the impressed frequency. It is clear, therefore that the velocity of light in the medium varies with the frequency of light.

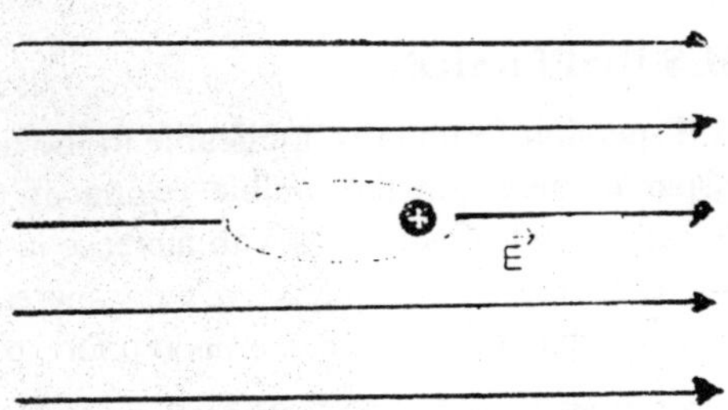

(a) : Displacement of electron cloud in a static electric field.

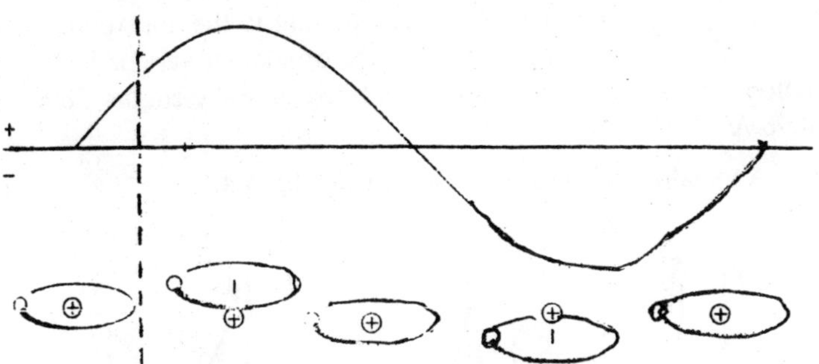

(b) : Oscillating atomic dipole.

Fig. 4.3

Also, refractive index depends upon the velocity of light in the medium. Therefore, refractive index of the medium varies with frequency (wavelength) of light.

DISPERSIVE POWER

According to Cauchy's formula $\mu = A + \frac{B}{\lambda^2}$ ...(1)

Differentiating equation (1), we get

$$\frac{d\mu}{d\lambda} = -\frac{2B}{\lambda^3} \quad ...(2)$$

$\frac{d\mu}{d\lambda}$ is the dispersive power of the medium. Therefore, dispersive power is inversely proportional to cube of the wavelength of light. At $\lambda = 4000$Å (violet) the dispersive power is about eight times the dispersive power at $\lambda = 8000$Å (red). It means that the spectral lines are more dispersed near the violet end of the spectrum than that at the red end.

ANOMALOUS DISPERSION

Cauchy's dispersion formula is applicable to the normal dispersion of many transparent media in the visible region of the spectrum. In general, the refractive index decreases with increase in wavelength. But, over small wavelength regions, the refractive index increases with increase in wavelength accompanied by increase absorption of the radiation.

Increase in refractive index with increase in wavelength is termed as *anomalous dispersion.* It is noticed that anomalous dispersion is present at those wavelengths corresponding to the absorption bands of the medium. For example, in the case of sodium vapour in the visible region, anomalous dispersion is noticed at wavelengths 5890 A and 5896 Å.

Anomalous dispersion is shown in Fig. 4.4.

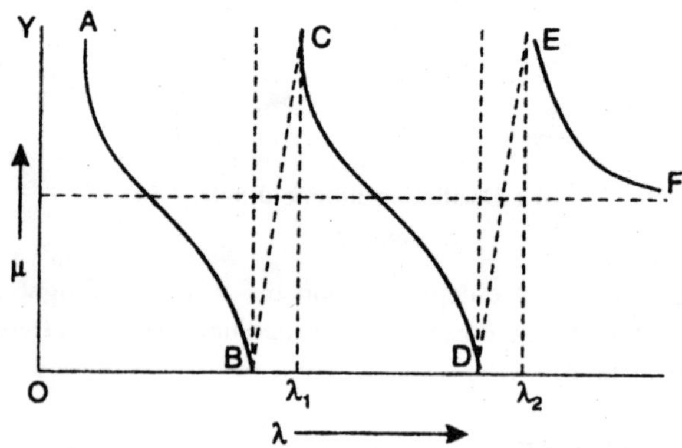

Fig. 4.4

The curves, AB, CD, and EP show normal dispersion. However, at the absorption wavelengths λ_1 and λ_2 the refractive index increases with wavelength. BC and DE represent anomalous dispersion. Iodine vapour has an absorption band in the visible region. When a hollow prism containing iodine vapour is used, red rays are deviated more than violet rays.

Sellmeier's formula for the refractive index of a medium is given by,

$$\mu^2 = 1 + \sum_p \frac{A_p \lambda^2}{\lambda^2 - \lambda_p^2} \qquad ...(1)$$

Here A_p is proportional to the number of particles per unit volume vibrating with a natural frequency corresponding to the wavelength λ_p and λ is the wavelength of incident radiation. If the medium has two absorption wavelengths corresponding to λ_o and λ_1 the Sellmeier's relation will be,

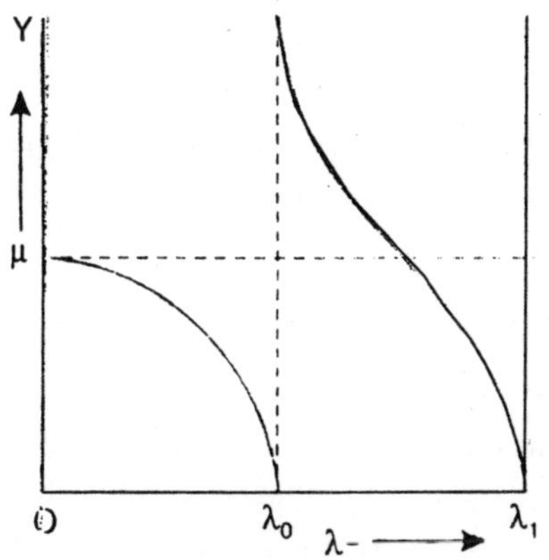

Fig. 4.5

$$\mu^2 = 1 + \frac{A_0 \lambda^2}{\lambda^2 - l_o^2} + \frac{A_1 \lambda^2}{\lambda^2 - l_o^2} \qquad ...(2)$$

From this relation it is clear that as μ approaches λ_o or λ_1 from the shorter wavelength side, μ tends to $-\infty$ at resonance. Similarly from the longer wavelength side μ tends to $+\infty$ at resonance. The curves obtained between μ and K are shown in Fig. 4.5.

Cauchy's dispersion formula is applicable for normal dispersion where as Sellmeier's formula explains both normal and anomalous dispersion.

CONSTITUTIVE RELATIONS

The electric and magnetic properties of a medium are described by three quantities: relative permittivity, ε, the relative permeability, μ, and the conductivity σ. If Maxwell's equations are to be extended to material media, they should be supplemented with relations which would contain these quantities characterizing individual properties of the medium. Such relations are called constitutive relations and also known as material equations.

The *permittivity* of a dielectric material is denoted by

$$\varepsilon = \varepsilon_0 \varepsilon_r \text{ F/m} \qquad ...(1)$$

where ε_r is a dimensionless quantity called the *relative permittivity* or *dielectric constant* of the material. ε_o is the *permittivity of free space* which is given by

$$\varepsilon_o = 8.854 \times 10^{-12} \text{ F/m}$$

In addition to the electric *field intensity* E, we often use the related quantity, the *electric flux density* D.

$$D = \varepsilon E \text{ C/m}^2 \qquad ...(2)$$

Magnetic *permeability* of a material is denoted by

$$\mu = \mu_0 \mu_r \text{ H/m} \qquad ...(3)$$

where μ_r is a dimensionless quantity called the *relative permeability* of the material and μ_o is called the magnetic permeability of free space.

$$\mu_o = 4\pi \times 10^{-7} \text{ henry/m}$$

Magnetic field intensity is denoted by H and is related to magnetic induction B through

$$B = \mu H \qquad ...(4)$$

The point form of Ohm's law describes the flow of current in a material and it is given by

$$J = \sigma E \qquad ...(5)$$

where J is the current density and σ the electrical conductivity of the medium.

We usually assume ε_r, μ_r, and σ are not functions of time and that the medium is linear, homogeneous, isotropic; ε, μ_r, and σ are therefore constant and uniform throughout the medium. A medium is a *homogeneous medium* when the quantities ε, μ, and σ are constant throughout the medium. The medium is *isotropic* if ε is a scalar constant, so that D and E have everywhere the same direction. Material equations have the simplest form for sufficiently weak electric fields. Equations (2), (4), and (5) hold good for the case of isotropic non-ferroelectric and non-ferromagnetic media.

When the relations (2), (4), and (5) are inserted in the Maxwell's equations, we get the following differential equations relating the electric and magnetic field strengths E and H. If they are then solved as simultaneous equations, they will determine the laws which both E and H must obey.

$$\nabla . E \frac{\rho}{\varepsilon} \quad \text{...(6)}$$

$$\nabla \times E = -\mu \frac{\partial H}{\partial t} \quad \text{...(7)}$$

$$\nabla . H = 0 \quad \text{...(8)}$$

$$\nabla \times H = J + \varepsilon \frac{\partial E}{\partial t} \quad \text{...(9)}$$

UNIFORM PLANE WAVES

When energy is emitted by a source, it expands outwardly from the source in the form of spherical waves. The spherical wave travels at the same speed in all directions and therefore expands at the same rate. To an observer very far away from the source, the wave front of the spherical wave appears approximately planar. A plane wave is the simplest example of wave motion. In a plane wave, the electric and magnetic intensities are of constant value over any plane perpendicular to the direction of propagation. Such a plane is a surface of equal phase. A *plane wave* is thus a wave for which the phase has the same value at all points on an infinite plane. As the amplitude is also constant over the plane surface, it is called a *uniform plane wave*. Plane waves vary only in the direction of propagation and are uniform in planes normal to the direction of propagation. In this chapter we confine ourselves to the plane wave propagation in unbounded media. Plane wave propagation can be described by Cartesian coordinates, which are easier to work with mathematically than the spherical coordinates needed for describing propagation of a spherical wave.

WAVE POLARIZATION

The polarization of an electromagnetic wave is taken as the direction of the E vector. It describes the *shape and locus of the tip of the E vector* (in the plane orthogonal to the direction of propagation) *at a given point in space as function of time.* The uniform plane wave described by has only y-component of electric field. The direction of the field vector at some point in space and time lies along a line in a plane perpendicular to the direction of propagation.

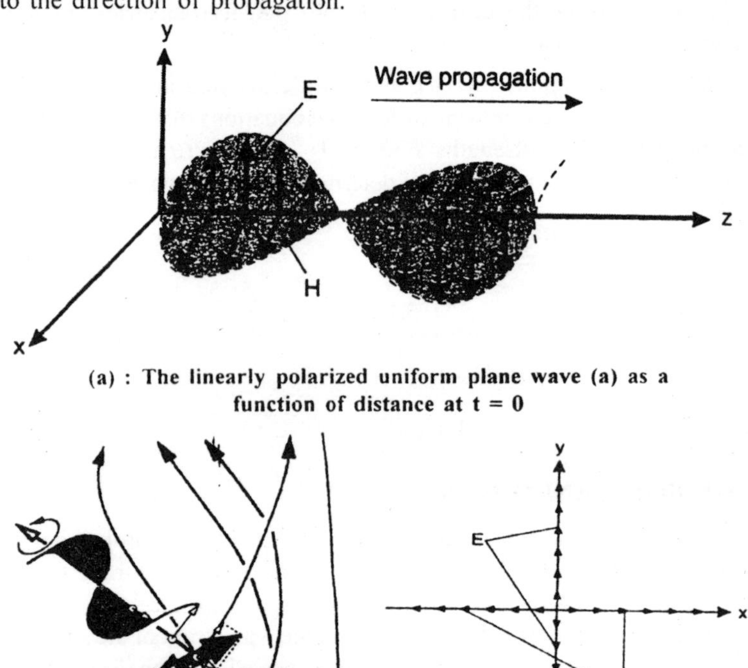

(a) : The linearly polarized uniform plane wave (a) as a function of distance at t = 0

(b) : As a function of time at z = 0.

Fig. 4.6

If we take a cross-sectional view of the wave in the xy-plane at z = 0, we observe the magnitude of the field vectors varying sinusoidally, as shown in Fig. 4.6 (b). Note that the field vectors reverse direction with each half-cycle but that this variation is always along a line (the ± y direction for E and the ± x direction for H). Since the direction of the field vectors at some point in space and every point in time lies along a line in a plane perpendicular to the direction of propagation, the wave is said to be linearly polarized. Thus, the uniform plane waves are *linearly polarized.*

In general, waves can have either E_y or E_x components when travelling in z-direction. If a wave has E_y-component, then the wave is polarized along y-direction and if the wave has E_x -component, then it is polarized along x-direction. If two coherent and linearly polarized waves polarized in mutually perpendicular directions propagate in the same direction, they combine to give a resultant wave whose state of polarization depends on their phase difference.

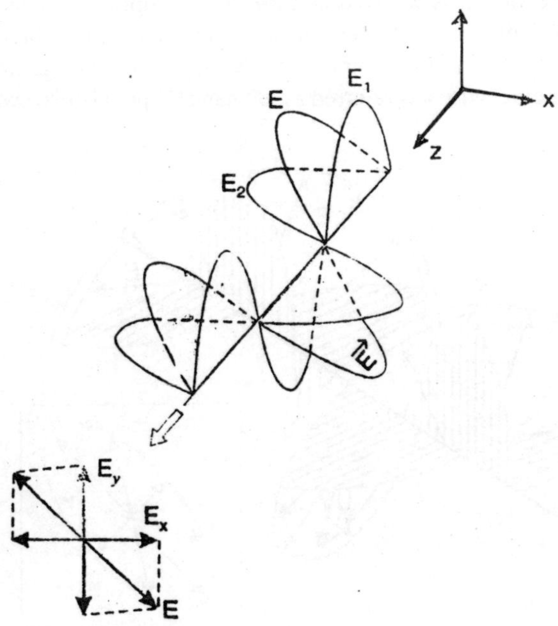

Fig. 4.7

Case 1 : If the waves are of equal or of unequal intensity ($E_y = E_x$ or $E_y \neq E_x$) and are in-phase, they may be represented by

$$E_1 = E_y \sin(\omega t - k z)$$

and
$$E_2 = E_x \sin(\omega t - k z)$$

The waves are shown in Fig. 4.7. These two waves may be considered as the rectangular components of a wave whose electric intensity is

$$E = iE_x + jE_y \qquad ...(1)$$

They combine to yield a linearly polarized wave, as shown in Fig. 4.7.

Case 2 : The waves may be of unequal intensity ($E_y \neq E_x$) and have a phase difference of 90°, as shown in Fig. 4.8. It is seen that one wave is equal to zero-at any given time and place, where the other wave is a maximum.

At any given place, as time is varied, the tip of the E vector traces an ellipse (see inset in Fig. 4.8) during the course of one cycle and the resultant wave is *elliptical polarized.* At any given instant of time the E vector describes an elliptical helix in space (Fig. 4.8). If E vector rotates in clockwise direction it is referred as right hand elliptically polarized wave since if one places the fingers of the right hand in the direction of the rotation, the thumb will point in the direction of propagation. If E rotates counter clockwise it is referred as left hand elliptical polarization.

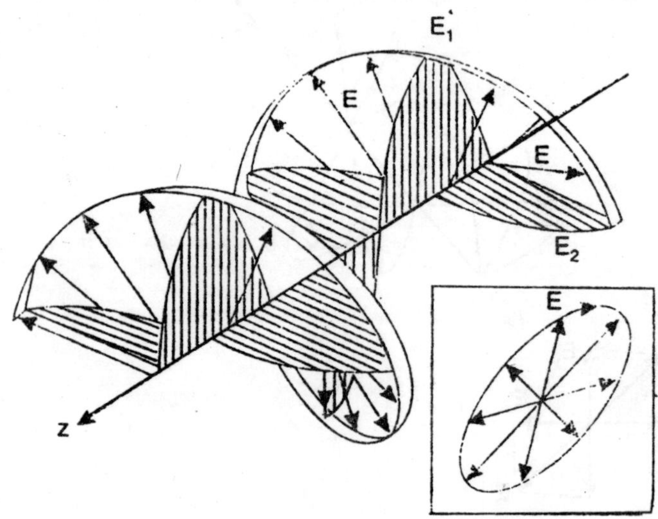

Fig. 4.8

Case 3 : In the particular case where the waves have equal intensity and maintain a phase difference of 90°, the tip of the E vector of the resultant wave traces out a circle and the wave is referred to as *circularly polarized* wave. At a given instant E and H describe the form of regular helixes, the two being everywhere 90° apart. If E vector rotates in clockwise direction it is referred *as right hand circularly polarized* wave and if it rotates counter clockwise it is referred as *left hand circular polarization.*

LIGHT WAVES AT BOUNDARIES

We have studied so far the propagation of a light wave in free space. When light proceeds from one medium into another (*e.g.*, from air into a solid substance), several things happen. Some of the light radiation may be transmitted through the medium, some is absorbed, and some is reflected at the interface between the two media. Obviously, the intensity I_o of the beam incident on the surface of the medium must equal the sum of the intensities of the transmitted, absorbed and reflected beams. Thus,

$$I_o = I_T + I_A + I_R \qquad ...(1)$$

where T_T, I_A and I_R are the intensities of the transmitted, absorbed and reflected beams respectively. Eqn. (1) may be rewritten as

$$1 = \frac{I_T}{I_o} + \frac{I_A}{I_o} + \frac{I_R}{I_o}$$

or $$T + A + R = 1 \qquad ...(2)$$

where $T = \frac{I_T}{I_o}$ is the fraction of the light transmitted by the material and is called the *transmitivity* or *transmittance*, $A = \frac{I_A}{I_o}$ is the fraction of light absorbed by the material and is called *absorptivity*, and $R = \frac{I_R}{I_o}$ is the fraction of light reflected by the material and is called the *reflectivity* or *reflectance* of the material. Glass and other transparent materials reflect only a small portion of the incident radiation and transmit the remainder. Absorption of light in the medium will be negligible. Therefore, we consider here only reflection and transmission of light. In such cases eqn. (2) reduces to

$$T + R = 1 \qquad ...(3)$$

We now develop the familiar laws of optics by means of Maxwell's equations.

Boundary Conditions

Maxwell's equations in the differential form presume that all quantities vary continuously in space and time. When the waves appear at the interface between two different materials, Maxwell's equations require that certain relationships must exist between the fields on the two sides of the interface. These are known as *boundary conditions*. They are obtained as follows.

At the boundary between two different media, we resolve the total electric field E into a component tangent to the boundary surface, E_t and a component normal to the boundary surface, E_n, as shown in Fig. 4.9 (a).

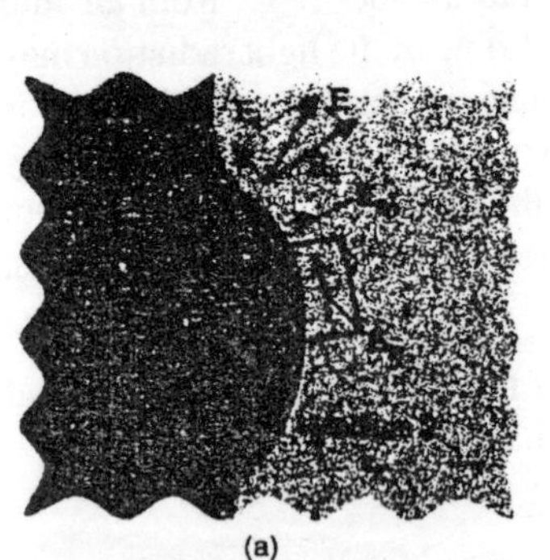
(a)

Electric field components at a boundary.

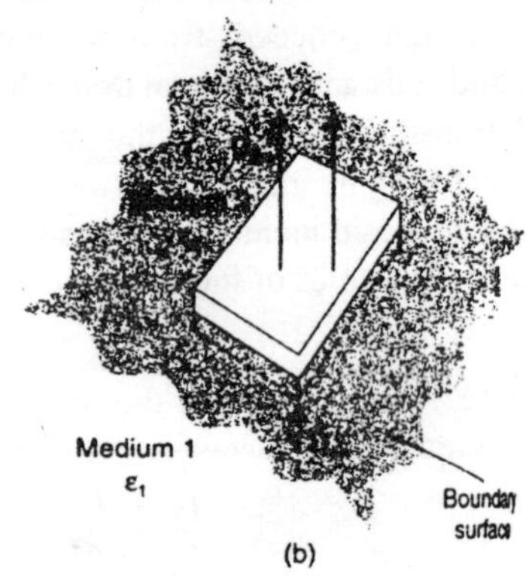

(b)

Fig. 4.9

1. The components of the electric field vector E that are tangent to the surface of a boundary must be continuous across the boundary. It means that the tangential components of E are equal on the two sides of the interface.

$$E_{2t} = E_{1t} \quad ...(4)$$

2. The difference between the normal components of D is equal to the free surface charge density at the boundary.

$$D_{2n} - D_{1n} = \rho_s \quad ...(5)$$

Again at the boundary between two different media, we resolve the total magnetic flux density B into a component tangential to the boundary surface, B_t and a component normal to the boundary surface, B_n, as shown in Fig. 4.10.

3. At the interface between two different materials, the normal component of B is continuous across the interface. It means that

the normal components of B are equal on the two sides of the interface.

$$B_{2n} = B_{1n} \quad ...(6)$$

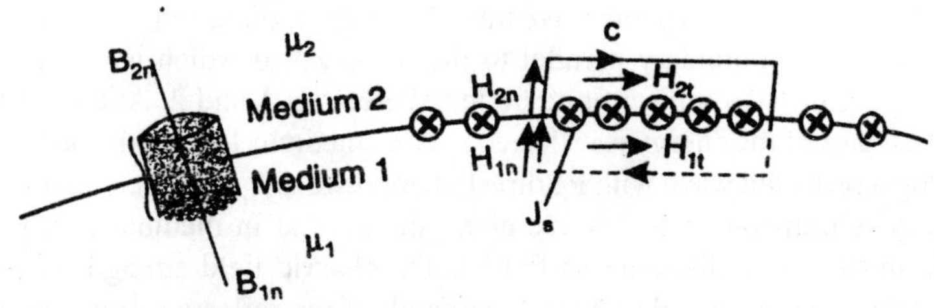

Fig. 4.10 : Magnetic field components at a boundary.

4. The difference between the tangential components of H is equal to the current density.

$$H_{2t} - H_{1t} = J_s \quad ...(7)$$

The boundary conditions at the interface between two dielectric materials, where there is no free charge and no current, are;

1. Tangential components of E and H are continuous across the interface. That is

$$E_{2t} = E_{1t} \quad ...(8)$$

$$H_{2t} = H_{1t} \quad ...(9)$$

2. Normal components of D and B are continuous across the interface.

$$D_{2n} = D_{1n} \quad ...(10)$$

$$B_{2n} = B_{1n} \quad ...(11)$$

These boundary conditions are used to write equations relating the fields on each side of the interface.

WAVE INCIDENT NORMALLY ON BOUNDARY

Let us consider a plane surface separating two unlike media, for instance air and glass. Since the dielectric constant of one material is different from that of the other, the same equations cannot describe the wave on the two sides of the boundary. Accordingly, an additional expression representing a reflected wave must be added. Thus, three distinct wave systems exist at every boundary of media. These three are known as the *incident*, the *reflected* and the *refracted waves*. It is our problem to find relations between the amplitude and phase of these three waves which satisfy the general boundary conditions.

Fig. 4.11 shows a plane wave traveling in the z-direction and incident *normally* on a boundary parallel to the plane z = 0 which is the plane separating the two semi-infinite perfect dielectrics 1 and 2. At the plane z = 0, a part of the energy is reflected back in medium 1 along the normal, giving a reflected wave with P_r. directed opposite to P_i, and the remaining energy is transmitted as a wave along the normal in medium 2. Thus, P_i is in the same direction as E_i.Ei is the electric field strength of the incident wave, E_r the electric field strength of the reflected wave and E_t the electric field strength of the transmitted wave propagated in the second medium. Similarly, H_1, H_r, and H_t represent the strengths of incident, reflected and transmitted magnetic fields. These three waves have to satisfy the conditions of continuity for the tangential components of E_i and H_r at the interface. Accordingly, the sum of the tangential components of E_i and E_r just to the left of the interface must be equal to the tangential component of E_t just to the right of the interface; a similar situation holds for H. Thus,

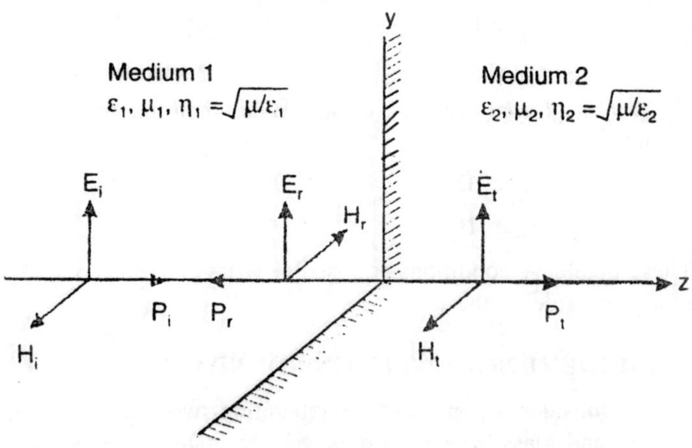

Fig. 4.11 : The boundary between two dielectric media is the xy plane. Medium 1 contains the incident and reflected waves, medium 2 the transmitted wave.

$$\left.\begin{array}{l}\text{Tangential component of electric field}\\ \text{in medium 1}\end{array}\right\} = \left\{\begin{array}{l}\text{Tangential component of electric field}\\ \text{in medium 2}\end{array}\right.$$

That is, $$E_i + R_r = E_t \qquad ...(12)$$

Similarly, applying the boundary condition for magnetic fields and noting that $\sigma = 0$ for perfect dielectrics, we get

$$H_i + H_r = H_t \qquad ...(13)$$

Also we have $$E_i = \eta_1 H_i \qquad ...(14)$$

$$E_r = -\eta_1 H_r \qquad ...(15)$$

and $$E_t = \eta_2 H_t \qquad ...(16)$$

Substituting these values, we get

$$\therefore \quad \frac{1}{\eta_1}(E_i - E_r) = \frac{1}{\eta_2}(E_i + E_r)$$

or $$\eta_2 (E_i - E_r) = \eta_1 (E_i + E_r)$$

or $$E_i (\eta_2 - \eta_1) = E_r (\eta_2 + \eta_1)$$

Therefore, the *reflection coefficient* is given by

$$\Gamma = \frac{E_r}{E_i} = \frac{\eta_2 - \eta_1}{\eta_2 + \eta_1} \qquad ...(17)$$

The *transmission coefficient* is given by

$$\tau = \frac{E_t}{E_i} = \frac{E_i - E_r}{E_i} = 1 + \frac{E_r}{E_i}$$

$$= 1 + \frac{\eta_2 - \eta_1}{\eta_2 + \eta_1}$$

or $$\tau = \frac{2\eta}{\eta_2 + \eta_1} \qquad ...(18)$$

If $\Gamma = 0$, it follows from eqn. (17) that there will be no reflection of the wave. This implies that $\Gamma_2 = \eta_1$ or there is no change of medium. From eqn. (17) and (18), it can be seen that Γ and τ are interrelated by the formula

$$\tau = 1 + r \qquad ...(19)$$

For nonmagnetic media $\eta_1 = \sqrt{\frac{\mu_1}{\varepsilon_1}}$ and $\eta_2 = \sqrt{\frac{\mu_2}{\varepsilon_2}}$ and using these values into (17) and (18), we respectively get

$$\Gamma = \frac{\sqrt{\varepsilon_1} - \sqrt{\varepsilon_2}}{\sqrt{\varepsilon_1} + \sqrt{\varepsilon_2}} = \frac{n_1 - n_2}{n_1 + n_2} \quad ...(20)$$

and

$$\tau = \frac{2\sqrt{\varepsilon_1}}{\sqrt{\varepsilon_1} + \sqrt{\varepsilon_2}} = \frac{2n_1}{n_1 + n_2} \quad ...(21)$$

The reflection and transmission coefficients are useful in finding the fields. Knowing the magnitude and direction of the E vector and intrinsic impedance of the material, the magnitude and direction of the H vector can be found.

SOLVED EXAMPLES

Example 1:

The wave function of a light wave is

$$E\ (z, t) = 10^3 \sin \pi\ (3 \times 10^6 x - 9 \times 10^{14} t)$$

(a) determine the speed, wavelength, frequency and period of the wave,

(b) determine the magnetic field associated with the wave.

Solution:

(i) $E\ (z, t) = 10^3 \sin p\ (3 \times 10^6 x - 9 \times 10^{14} t)$

$$= 10^3 \sin 3 \times 10^6 \pi\ (x - 3 \times 10^8 t) \quad ...(1)$$

The equations is similar to the general wave equation

$$E\ (z, t) = E_0 \sin k\ (x - vt) \quad ...(2)$$

Comparing (1) with (2), we get,

$$v = 3 \times 10^8 \text{ m/s, and } k = 3 \times 10^6 \pi/\text{m}$$

$$\therefore \quad \lambda = \frac{2\pi}{k} = \frac{2\pi}{3 \times 10^6 \pi} = 6.666 \times 10^{-7}\text{m} = 6666 \text{ Å}$$

$$\nu = \frac{v}{\lambda} = \frac{3 \times 10^8 \text{ m/s}}{6.666 \times 10^{-7} \text{ m}}$$

$$= 4.5 \times 10^{14} \text{ Hz}$$

$$T = \frac{1}{v} = \frac{1}{4.5\times10^{14}\,\text{Hz}} = 2.2 \times 10^{-15}\text{s}$$

(ii) The light wave propagates in the z-direction while the E-vector oscillates along x-direction in xz-plane. Since in a EM wave, the magnetic field B is normal to both E and wave propagation directions, it is should be in the yz-plane.

Thus, $B_x = 0;\ B_z = 0$

$$B = B_y\,(z, t) = B\,(z, t)$$

As $E = cB,\ B = \frac{E}{c}$

$$\therefore\ B(z, t) = \frac{10^3 \sin\pi(3\times10^6 y - 9\times10^{14} t)}{c}$$

$$= \frac{10^3 \sin\pi(3\times10^6 y - 9\times10^{14} t)}{3\times10^8}$$

$$= 3.33 \times 10^{-6} \sin\pi\,(3 \times 10^6 y - 9 \times 10^{14} t)$$

Example 2:

A electromagnetic wave moving though free space has an electric field given by,

$$E = 100 \sin\left[8\pi\times10^{14}\left(t - \frac{z}{3\times10^8}\right)\right].$$

Calculate the corresponding intensity.

Solution:

$$E = 100 \sin\left[8\pi\times10^{14}\left(t - \frac{z}{3\times10^8}\right)\right];\ \therefore\ E_0 = 100\ \text{V/m}$$

$$\text{Intensity, } I = \left(\frac{c\varepsilon_0}{2}\right)E_0^2$$

$$= \frac{(3\times10^8)\left(8.85\times10^{-12}\,\frac{C^2}{Nm^2}\right)(100\,\text{V/m})^2}{2}$$

Example 3:

Determine the reflection coefficient and the transmission coefficient of an electric field wave traveling in air and incident normally on a boundary between air and dielectric having permeability = μ_0 and permittivity, $\varepsilon_r = 4$.

Solution:

We know that, $\eta_1 = \sqrt{\dfrac{\mu_0}{\varepsilon_0}}$

and $\eta_2 = \sqrt{\dfrac{\mu_0}{\varepsilon_0 \varepsilon_r}}$

But it is given that, $\varepsilon_r = 4$. $\therefore\ \eta_2 = \dfrac{1}{2}\sqrt{\dfrac{\mu_0}{\varepsilon_0}} = \dfrac{1}{2}\eta_1$

Reflection coefficient is given by,

$$\Gamma = \frac{\eta_2 - \eta_1}{\eta_2 + \eta_1}.$$

Putting the values of η_1 and η_2, we have,

$$\Gamma = \frac{\dfrac{\eta_2}{2} - \eta_1}{\dfrac{\eta_1}{2} + \eta_1} = -\frac{1}{3} = -0.33$$

Transmission coefficient is given by,

$$\tau = \frac{2\eta_2}{\eta_2 + \eta_1} = \frac{2 \times \dfrac{\eta_1}{2}}{\dfrac{\eta_1}{2} + \eta_1} = \frac{2}{3} = 0.66$$

Example 4:

A uniform plane electromagnetic wave is incident at angle θ_I at the surface of discontinuity between two homogenous isotropic dielectrics with permittivity ε_I and ε_2. ε_2 being the electric intensities respectively of the incident, reflected and transmitted waves, show that the reflection coefficient for parallel polarization is given by,

$$\frac{E_r}{E_i} = \frac{\tan(\theta_1 - \theta_2)}{\tan(\theta_1 + \theta_2)}$$

Solution:

We know that, $\dfrac{E_r}{E_i} = \dfrac{\sqrt{\varepsilon_2}\cos\theta_1 - \sqrt{\varepsilon_1}\cos\theta_2}{\sqrt{\varepsilon_2}\cos\theta_1 + \sqrt{\varepsilon_1}\cos\theta_2}$

Dividing numerator and denominator by $\sqrt{\varepsilon_2}$,

we get,

$$\frac{E_r}{E_i} = \frac{\cos\theta_1 - \sqrt{\frac{\varepsilon_1}{\varepsilon_2}}\cos\theta_2}{\cos\theta_1 + \sqrt{\frac{\varepsilon_1}{\varepsilon_2}}\cos\theta_2}$$

$$= \frac{\cos\theta_1 - \frac{\sin\theta_1}{\sin\theta_2}\cos\theta_2}{\cos\theta_1 + \frac{\sin\theta_1}{\sin\theta_2}\cos\theta_2}$$

$$= \frac{\sin\theta_2\cos\theta_1 - \sin\theta_1\cos\theta_2}{\cos\theta_1\sin\theta_2 + \sin\theta_1\cos\theta_2}$$

Dividing again numerator and denominator by cos θ_1, and cos θ_2, we get,

$$\frac{E_r}{E_i} = \frac{\tan\theta_2 - \tan\theta_1}{\tan\theta_2 + \tan\theta_1}$$

Dividing again numerator and denominator by 1 + (tan θ_1 tan θ_2) we get,

$$\frac{E_r}{E_i} = \frac{\tan(\theta_2 - \theta_1)}{(\theta_2 + \theta_1)}; \text{ Hence proved.}$$

EXERCISES

1. Establish Maxwell's equations for electromagnetic fields and obtain an expression for pointing vector.
2. Give an account of electromagnetic theory of dispersion.
3. Find the velocity of a plane wave in a lossless medium having a relative permittivity of 5 and relative permeability of 1.
4. A plane sinusoidal linearly polarized electromagnetic wave of wavelength $\lambda = 5 \times 10^{-7}$m travels in vacuum in the direction of the X-axis. The average intensity of the wave per unit area is 0.1 W/m^2 and plane of the vibration of the electric field is parallel to the Y-axis. Write the equations describing the electric and magnetic fields of the wave.
5. In free space, H(z, t = 1.33×10^{-1} cos ($4 \times 10^7 - \beta z$) a_x A/m. Obtain an expression for E(z, t). Find β and λ.
6. State and explain Maxwell's equations for electromagnetic field. Starting from Maxwell's equations, deduce the wave equation for a plane wave in free space.

7. Show that the ratio of the electric and magnetic fields of a uniform plane wave is constant depending upon the medium.
8. What is meant by the Poynting vector? Derive Poynting vector from Maxwell's equations and explain its physical significance.
9. Solve the Maxwell's equations for perfect dielectric containing no charge and no conduction current.
10. What must be the strength of uniform electric field if it is to have the same energy density as that possessed by a 400 gauss magnetic field [1 W/m^2 10^4 ausses, ε_0 = 8.85 × 10^{12} F/m and μ_0 = 4π × 10^{-7} H/m].
11. Find the Brewster angle for a parallel-polarized wave traveling from air into glass for which ε_r = 5.0
12. In free space, E(z, t) =1.0 sin (ωt – βz)a_x V/m. Show that the average power crossing a circular disk of radius 15.5 m in a z = const. plane is 1 W.